Essentials of Clinical Neurophysiology

Essentials of Clinical Neurophysiology

Third Edition

Karl E. Misulis, M.D., Ph.D.
Clinical Professor of Neurology, Vanderbilt University School of Medicine, Nashville, Tennessee and University of Tennessee College of Medicine, Memphis; Neurologist, Semmes-Murphey Neurologic and Spine Institute, Jackson, Tennessee

Thomas C. Head, M.D.
Neurologist, Semmes-Murphey Neurologic and Spine Institute, Jackson, Tennessee

An imprint of Elsevier Science

An imprint of Elsevier Science

200 Wheeler Road
Burlington, MA 01803

ESSENTIALS OF CLINICAL NEUROPHYSIOLOGY   ISBN 0-7506-7441-5
Copyright 2003, Elsevier Science (USA). All rights reserved.

No part of this publication may be reproduced or transmitted in any form or by any means, electronic or mechanical, including photocopy, recording, or any information storage and retrieval system, without permission in writing from the publisher (Butterworth-Heinemann, 200 Wheeler Road, MA 01803).

NOTICE

Medicine is an ever-changing field. Standard safety precautions must be followed but as new research and clinical experience broaden our knowledge, changes in treatment and drug therapy may become necessary or appropriate. Readers are advised to check the most current product information provided by the manufacturer of each drug to be administered to verify the recommended dose, the method and duration of administration, and contraindications. It is the responsibility of the treating physician, relying on experience and knowledge of the patient, to determine dosages and the best treatment for each individual patient. Neither the Publisher nor the author assumes any liability for any injury and/or damage to persons or property arising from this publication.

The Publisher

Previous editions copyright 1997, 1993

Library of Congress Cataloging-in-Publication Data

Misulis, Karl E.
  Essentials of clinical neurophysiology / Karl E. Misulis, Thomas C. Head.–3rd ed.
    p.; cm
  Includes bibliographical references and index.
  ISBN 0-7506-7441-5
    1. Electroencephalography.   2. Electromyography.   3. Nervous system—Diseases—Diagnosis   4. Evoked potentials (Electrophysiology).   5. Neurophysiology.   I. Head, Thomas C. (Thomas Channing), 1963-   II. Title.
    [DNLM: 1. Electroencephalography–methods.   2. Electromyography–methods.   3. Evoked Potentials–physiology   4. Neural Conduction–physiology.   5. Polysomnograhy–methods.
WL 102   M678e   2003]
RC386.6.E43   M57   2003
616.8′047547–dc21   2002026299

Acquisitions Editor: Susan F. Pioli

CE/MVY

Printed in the United States of America

Last digit is the print number: 9 8 7 6 5 4 3 2 1

Contents

Preface	vii
Part I: Basic Physiology and Physics	**1**
1. Nerve and Muscle Physiology	3
2. Physics and Electronics	15
Part II: Electroencephalography	**39**
3. Electroencephalography Basics	41
4. Normal Electroencephalographic Patterns	65
5. Abnormal Electroencephalographic Patterns	85
6. Neonatal Electroencephaolography	107
7. Special Studies in Electroencephalography	117
Part III: Nerve Conduction Study and Electromyography	**127**
8. Basic Principles of Nerve Conduction Study and Electromyography	129
9. Approach to Clinical Questions	161

10. Neuropathies	167
11. Myopathies	181
12. Neuromuscular Transmission Defects	187
Part IV: Evoked Potentials	**193**
13. Evoked Potentials Basics	195
14. Visual-Evoked Potentials	201
15. Brainstem Auditory-Evoked Potentials	211
16. Somatosensory-Evoked Potentials	221
Part V: Polysomnography	**231**
17. Physiology of Sleep and Sleep Disorders	233
18. Sleep Disorders	239
Appendix: Abbreviations and Units of Measurement	249
Glossary	251
References	257
Index	259

Preface

Much has changed in the five years since the second edition of this text. Not only are there methodological changes but clinical correlations have been refined and expanded, and the purpose of neurophysiologic studies in clinical practice has changed. EEG and EMG have had improvements in digital technology for acquisition and analysis. EPs are still helpful as a functional study, even though structural studies, especially MRI, have replaced some of the purposes of EPs.

For the third edition, the text has been completely reorganized and rewritten. A second author has been added to help especially with the neuromuscular sections of the book. The companion CD includes some material removed from the original text that was felt not to be of prime importance to the student of clinical neurophysiology. In addition, the CD provides the opportunity to present some additional material including figures, video, and self-assessment questions not available in the printed text.

The target audience of this text continues to be students of clinical neurophysiology. This is not a reference text for the neurophysiology expert but can be a guide to good performance and interpretation of these studies. Comprehensive texts on the particular fields are listed in the References, and should be consulted.

We would like to express great appreciation for Susan F. Pioli and the staff at Elsevier Science, for their work on this and previous projects. We would also like to express our appreciation for all who have contributed information and feedback on this and previous editions of the text.

K.E.M.
T.C.H.

Part I Basic Physiology and Physics

1   Nerve and Muscle Physiology

Physiology of Excitable Tissue

Membranes are composed of lipid bilayers through which proteins are inserted. The extracellular and intracellular fluids are composed of ionic solutions. The positive ions (cations) are mainly sodium (Na^+), potassium (K^+), and calcium (Ca^{2+}). The anions are mainly chloride (Cl^-) and negatively charged proteins (which we will call $Prot^-$). The ionic fluids are good conductors of electricity, since they allow for free flow of the ions. However, the lipid bilayer is a good insulator, essentially making the extracellular and intracellular fluids separate conductors. The basic elements of membrane structure and function are presented in Figure 1.1.

Some of the membrane proteins are channels that allow certain ions to pass. Other proteins are pumps that actively push ions through the membranes against a concentration gradient. The sodium–potassium pump pushes Na^+ out of the cell while pushing K^+ into the cell. This creates a chemical gradient for Na^+ to go into the cell and K^+ out of the cell. Without the channel proteins, virtually none of the ions would move in or out of the cell; however, the channel proteins allow for controlled movement of the ions.

The channel proteins have very specific functions and characteristics. At rest, the potassium channel is open and the sodium and chloride channels are largely closed. During an action potential the sodium channel opens for a brief time then closes. Inhibitory stimuli produce opening of a channel, which allows both potassium and chloride to flow.

The next sections discuss how these functions produce the electrical properties of nerve and muscle.

Membrane Potentials

Differential permeability of cell membranes to ions produces a membrane potential. The sodium–potassium pump does not pump the cations exactly one-for-one, so there is a slight membrane potential created by this discrepancy. However, the largest component of the membrane potential is related to efflux of potassium from the cytoplasm to the extracellular space.

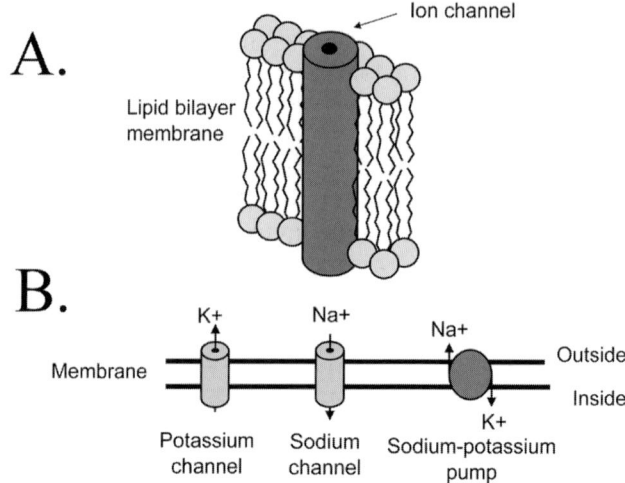

FIGURE 1.1 Membrane structure and function. **A.** Lipid bilayer membrane is impermeable to ions, but there are ion channels traversing the membrane. **B.** Diagrammatic representation of the lipid bilayer membrane with ionic channels that allow ions to move down electrochemical gradients, and ion pumps that establish the chemical gradients.

At rest, the permeability of the neuronal membrane to K^+ is greater than to any other ion. Therefore, K^+ diffuses out of the cell down its chemical gradient. Anionic proteins cannot accompany K^+ through the channel, so they are left behind. This results in the interior of the cell becoming negative compared to the exterior.

The potential being created across the cell membrane opposes further flow of K^+ down the chemical gradient. This electrical gradient opposes the flow of K^+ down the chemical gradient. When the electrical gradient is equal in strength to the chemical gradient, flow of K^+ ceases. The system is in equilibrium and this is termed the *equilibrium potential*. For K^+, the equilibrium potential is –75 mV.

Equilibrium Potential

The equilibrium potential is the potential at which the electrical gradient is sufficient to exactly counterbalance the chemical gradient.

The equilibrium potential for each ion can be calculated from the *Nernst equation*:

$$E_K = -58 \log \frac{[K]_i}{[K]_o}$$

$$E_{Na} = -58 \log \frac{[Na]_i}{[Na]_o}$$

$$E_{Cl} = -58 \log \frac{[Cl]_o}{[Cl]_i}$$

where E_K, E_{Na}, and E_{Cl} are the equilibrium potentials for sodium, potassium, and chloride, respectively. The bracketed letters are the concentrations of the ions with the following i and o representing the concentrations *inside* and *outside* of the cell, respectively. The concentrations for chloride are reversed, with the outside concentration over the inside concentration because of the negative charge of chloride. There are other ions, but these are the most important for the resting membrane potential.

The resting membrane potential is a complex summation of the potentials generated by each of the ions. The relative contribution of each ion to the resting membrane potential is dependent on the permeability of the membrane to that ion. This is described by the *Goldman constant field equation*:

$$V = -58 \log \frac{G_K[K]_i + G_{Na}[Na]_i + G_{Cl}[Cl]_o + \cdots}{G_K[K]_o + G_{Na}[Na]_o + G_{Cl}[Cl]_i + \cdots}$$

where G is the conductance of the membrane to the ion. Conductance is essentially permeability, although there is a semantic difference. Permeability refers to fluid dynamics, whereas conductance refers to flow of electric charge. Since this discussion is making the jump from chemistry to electricity, we will talk of conductance.

One could calculate the Goldman equation but in reality, there is such a difference in conductance between ions that it can be simplified. Since G_K is so much larger than G_{Na} and G_{Cl}, the potassium term takes over the equation. This means that the equation reduces to:

$$V = -58 \log \frac{G_K[K]_i}{G_K[K]_o}$$

Since the G_K drops out:

$$V = -58 \log \frac{[K]_i}{[K]_o}$$

This is essentially the same as the equilibrium potential for K^+. Therefore, the resting potential approaches -75 mV but is not exactly that value because of a small contribution from Cl^- and Na^+.

Generator Potentials

Generator potentials are produced by changes in conductance, so that the Goldman constant field equation results in a potential that differs from the equilibrium potential of potassium. The generator potential usually leads to an action potential, discussed below.

A generator potential can be produced by a variety of stimuli, including:

- Transmitter stimulation
- Electrotonic conduction from surrounding membrane
- Mechanical deformation
- Leaky membranes

Neurotransmitters are released from the presynaptic terminal and bind to receptors on the postsynaptic membrane. Binding results in an increased conductance to sodium and calcium that then depolarizes the membrane. This is the *generator potential*.

The Goldman constant field equation is no longer simplified when the conductance to these ions is increased. The equilibrium potential for sodium is positive, about +45 mV, so the membrane potential becomes more positive—meaning less negative. This is the depolarization of the *generator potential*.

Some neurotransmitters activate channels for potassium and chloride. Influx of chloride into the cell plus efflux of potassium out of the cell makes the membrane potential more close to the normal resting membrane potential. This acts to inhibit the effects of generator potentials, since an increase in sodium conductance is no longer able to produce the same degree of depolarization. While this is not a generator potential, the physiology of membrane channel conductance makes discussion here appropriate.

Electrotonic Conduction

Depolarization occurs at a site on the membrane with a certain geography. The depolarization affects adjacent membrane, however. If one records the membrane potential at sites progressively farther from a site of depolarization, the magnitude of the depolarization at that site will be less, as shown in Figure 1.2.

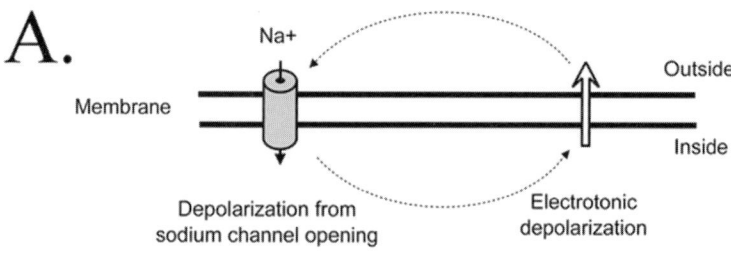

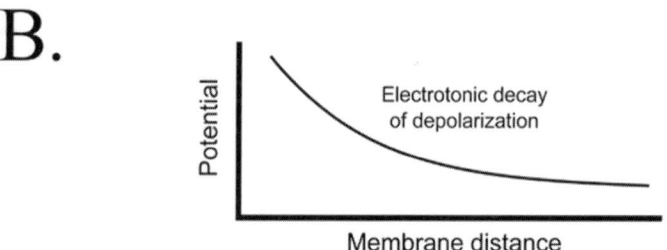

FIGURE 1.2   *Electrotonic conduction.* **A.** *Lipid bilayer membrane traversed by a sodium channel. Opening of this channel results in depolarization that is maximal at that site, but that is also present downstream from the site of channel opening.* **B.** *Graph of the depolarization as a function of distance from the site. The decay is exponential.*

Electrotonic conduction not only works within a cell, but can also work between cells if there is a tight junction that allows for electrotonic conduction from cell to cell. This is termed an *ephapse*. In the next cell, the depolarization may be sufficient to generate an action potential even if there was no action potential in the first cell.

Action Potentials

Action potentials are transient depolarizations that are sufficient to make the membrane potential positive. Not all neurons have action potentials; some conduct electronically, with depolarization of one region of the cell having an effect on depolarizing adjacent membrane without an action potential. All peripheral nerves and major relay nerves of the brain have action potentials. The phases of the action potential are:

- Generator potential
- Regenerative depolarization
- Peak potential
- Repolarization
- After-hyperpolarization

The *generator potential* was discussed in the previous section. The *regenerative depolarization* is due to activation of *voltage-gated* sodium channels. Figure 1.3 shows the sequence of changes in conductance which comprise the action potential. These channels are opened by the membrane potential reaching a certain voltage level that differs for each channel. If the potential threshold is crossed, the channel briefly opens. The increased sodium conductance produces further depolarization, and so on. Eventually, all of the

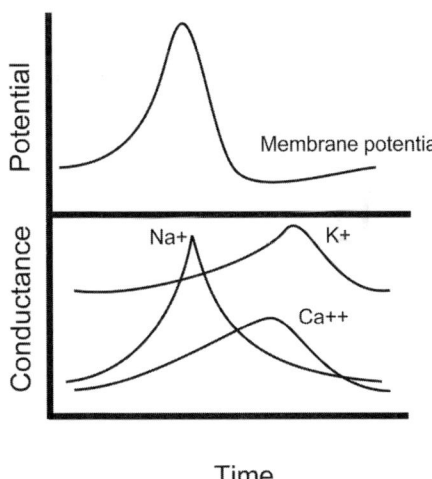

FIGURE 1.3 *Sequence of membrane potential and ionic conductance changes during an action potential. Potassium conductance predominates during rest; sodium conductance is markedly increased during the action potential.*

sodium channels that can open have opened, so the membrane potential reaches a *peak potential* that is near the equilibrium potential for sodium, about +45 mV.

Repolarization is accomplished by closing of the sodium channels and opening of potassium channels. The closing of the sodium channels is because they are *time dependent*, that is, they only have a fixed duration of opening. When the sodium conductance decreases, the Goldman constant field equation reverts to simplification such that the membrane potential is close to the equilibrium potential for potassium. This effect is augmented by opening of potassium channels. In fact, this opening results in transient hyperpolarization of the cell. This *after-hyperpolarization* makes it difficult for the cell to be excited again. This is part of the reason for a *refractory period*, where the cell cannot be stimulated to generate an action potential. The other reason for the refractory period is a delay in time before the voltage-gated sodium channels can be activated again.

Action Potential Propagation

An action potential develops in one part of the cell membrane and is conducted to adjacent membrane. The mechanism is similar to that of the generator potential with depolarization of membrane not involved in the action potential producing a regenerative opening of sodium channels and resultant action potential. This activates the next segment of nerve, and so on. The refractory period of the membrane that just sustained the action potential lessens the possibility of bidirectional conduction in the membrane. This is the mechanism of conduction in unmyelinated axons.

Myelinated axons conduct more quickly than unmyelinated axons because of the myelin sheath. The sheath serves several purposes, but the main purpose is to decrease the ionic leak of the axon. As shown in Figure 1.4,

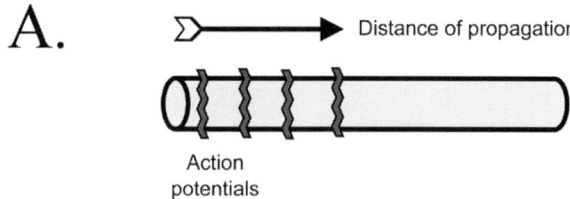

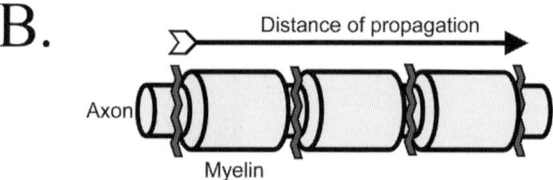

FIGURE 1.4 *Action potential propogation: myelinated vs. unmyelinated axon.* **A.** *Unmyelinated axon with propagation of the action potential down adjacent axonal membrane.* **B.** *Myelinated axon with propagation of the action potential from node to node, thereby conducting faster than in A.*

the depolarization is conducted farther down the axon when there is a myelin sheath than when the axon is unmyelinated. Since it takes time for the action potential to be generated, the myelination results in faster conduction down the axon.

Neurotransmitters

Transmitter Release

Action potentials or electrotonic conduction produce depolarization of the nerve terminal. This depolarization produces opening of ionic channels, including those that admit calcium. Calcium entry causes fusion of the vesicles with the membrane at the release site. There, the vesicles open up and dump transmitter into the synaptic cleft (Figure 1.5).

Not all neurons develop action potentials. If the axon has an action potential, a fixed amount of transmitter is released each time. If the neuron does not develop an action potential, then the amount of transmitter released is graded, dependent on the degree of depolarization of the terminal.

Receptor Binding

Transmitter diffuses across the narrow synaptic cleft and binds to the receptor for the transmitter on the postsynaptic membrane. Binding to the receptors

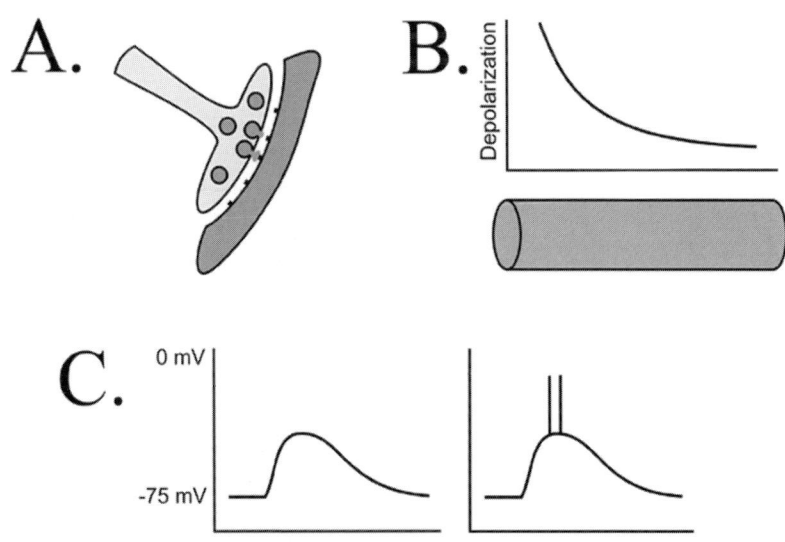

FIGURE 1.5 Synaptic transmission. **A.** Synaptic transmission with release of transmitter from the presynaptic terminal and binding of the transmitter to receptors on the postsynaptic membrane. **B.** Electrotonic conduction of depolarization from synaptic transmission down the postsynaptic membrane. **C.** Excitatory postsynaptic potential (EPSP) produced by synaptic transmission. The EPSP on the left did not reach threshold for action potential generation, but the one on the right did.

produces the effect of the transmitter. Typically, the receptor binding produces opening of membrane ionic channels.

Postsynaptic Potentials

Receptor activation produces *excitatory postsynaptic potentials* (EPSP) or *inhibitory postsynaptic potentials* (IPSP). EPSPs are due to increased conductance to sodium, calcium, or both. EPSPs sum to allow the adjacent membrane to reach threshold.

IPSPs are due to increased conductance to potassium and chloride. This clamps the membrane potential in the negative range, so that the membrane cannot reach threshold for action potential generation or even for ion influx.

Muscle Physiology

Neuromuscular Transmission

Neuromuscular transmission is essentially similar to synaptic transmission as described above. Action potential propagation down the motoneuron axon produces release of acetylcholine from the nerve terminal. The acetylcholine crosses the small gap between nerve and muscle and binds to the acetylcholine receptor. This opens channels in the muscle that allow for influx of sodium and calcium. These channels are more permissive than many other channels that allow specific ions. The movement of so many ions results in loss of the negative resting membrane potential, which in turn triggers an action potential in the adjacent muscle fiber membrane.

Electrical Properties of Muscle

Activation of the skeletal muscle results in action potentials that are propagated through the muscle fiber. Depolarization is conducted throughout the muscle fiber, which causes release of the calcium. The calcium, in turn, plays an important part in control of muscle contraction.

Muscle Contraction

Calcium and ATP are used to facilitate repeated cross-linking and releasing of the actin and myosin filaments, as shown in Figure 1.6. Calcium is essential for the cyclical binding and release of the filaments. Re-uptake of the calcium into the sarcoplasmic reticulum produces relaxation of the muscle.

Electrical Activity of Nerve

The generators of electroencephalography (EEG) and electromyography (EMG) are complex, and for most clinical practice the subcellular details are unimportant. Clinical correlations to particular electrophysiologic recordings are more important. Many purists conclude that interpretation of EEG, EMG, and evoked potentials (EP) is best if the interpreting physician makes judgments based on physiologic reasoning rather than pattern recognition. This is

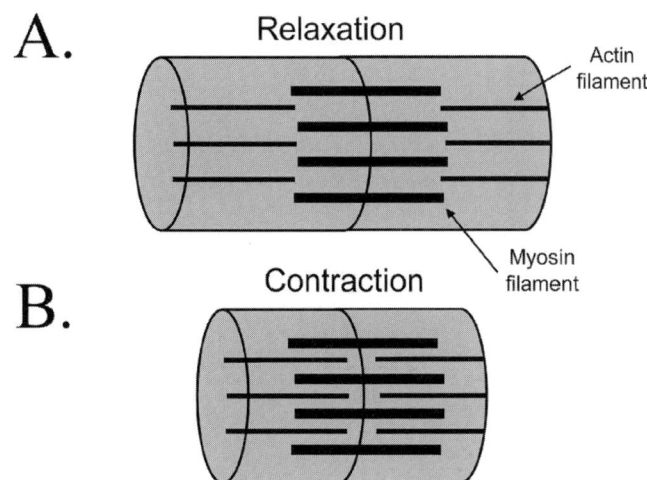

FIGURE 1.6 *Muscle contraction.* ***A.*** *Muscle fiber in the relaxed state.* ***B.*** *Muscle fiber contracted. Crosslinking between actin and myosin filaments shortens the muscle.*

often not practical, however, since some generators are very complex. Therefore, routine electrodiagnosic interpretation is a combination of pattern recognition and reasoning, where possible.

Most electrophysiologic recordings are of either nerve bundles or nerve connections. All recordings are extracellular, so intracellular recordings will not be considered.

Nerve Bundles

Nerve bundles in the peripheral or central nervous system conduct action potentials. For many studies, there are synchronous action potentials in a nerve bundle, such as in nerve conduction velocity measurements or EPs where the nerve was electrically stimulated in one location and recordings made from another location. Extracellular recordings, particularly made through the skin, are unable to detect a single action potential, so a synchronous volley is essential to detection.

Nerve Connections

Nerves have synapses mainly in the cerebral cortex, cerebellar cortex, and nuclei. In the cerebral cortex, the synapses are oriented throughout the layers. Major cortical efferent neurons are oriented vertically with dendritic arborizations such that excitatory or inhibitory activity can produce an electric field within the cortex that has a vertical axis and a radial distribution.

Most of the electric activity recorded during EEG is synaptic activity, whereas the signal recorded during measurement of nerve conduction is nerve conduction, that is, action potentials.

Scalp EEG recordings cannot "see" a single neuron, so activity has to be synchronous over many neurons before a wave is produced. Even then, it is attenuated through passage through skull and scalp. Seizure activity is discussed in Chapter 3, but typically differs from most non-epileptic activity by the synchrony of the electric potentials.

What We Record During Electrophysiologic Studies

Most potentials that we record during electrophysiological studies are *far-field potentials*, though some are *near-field* and some are mixed. The difference between these two types is not merely distance, as the names would suggest, but rather there are fundamental physical differences.

The influx of sodium into the cell during an action potential is effectively an inward current. An intracellular electrode would see this as a positive potential, because the interior is becoming more positive than it was at rest. An extracellular electrode would see this as a negative potential because of the shift of positive charge from the extracellular space into the cell. The extracellular potential can be recorded at a considerable distance from the cell and is known as a *field potential*. The movement of charge from excitable tissue through surrounding tissues is called *volume conduction*.

Near-Field Potentials

Near-field electrodes are recorded by electrodes close to the membrane, whereas far-field potentials are recorded at a distance. To illustrate, consider a peripheral nerve conducting a volley of action potentials. Refer to Figures 1.7 to 1.9 for this discussion. The near-field potential can be recorded by a unipolar or bipolar recording arrangement. For a unipolar recording, the active electrode (G1 for Grid 1 from old tube terminology) is directly

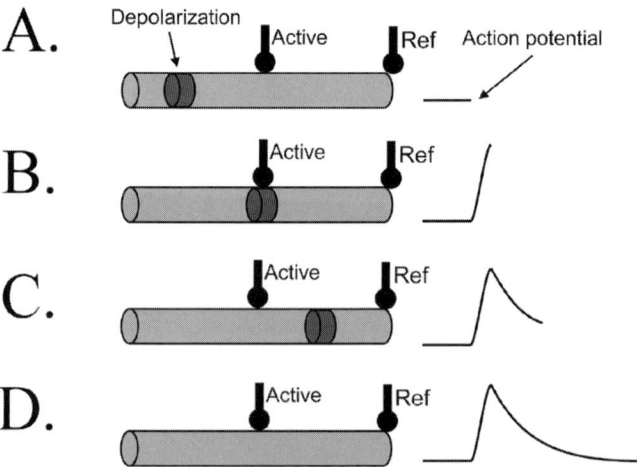

FIGURE 1.7 Unipolar near-field potential. **A.** The wave of depolarization has not reached the recording electrodes. **B.** The depolarization is maximal at the active electrode, the inward current producing a negative extracellular potential. **C.** The depolarization has passed the active electrode, so the potential is returning toward baseline. **D.** The depolarization has passed so the recorded potential is at baseline. The reference electrode is electrically inactive, since it is on a distant region.

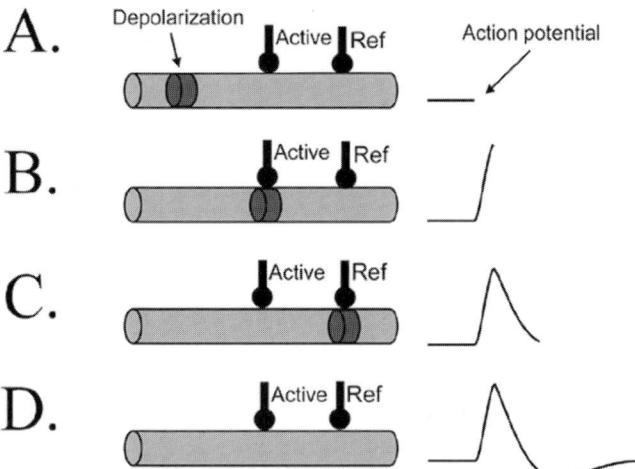

FIGURE 1.8 Bipolar near-field potential. **A.** The wave of depolarization has not reached the active electrode. **B.** Depolarization is maximal at the active electrode, producing a negative potential. **C.** Depolarization has passed the active electrode, so the potential declines. **D.** Depolarization is closest to the reference electrode, so a positive potential is seen. Eventually, the potential returns to baseline.

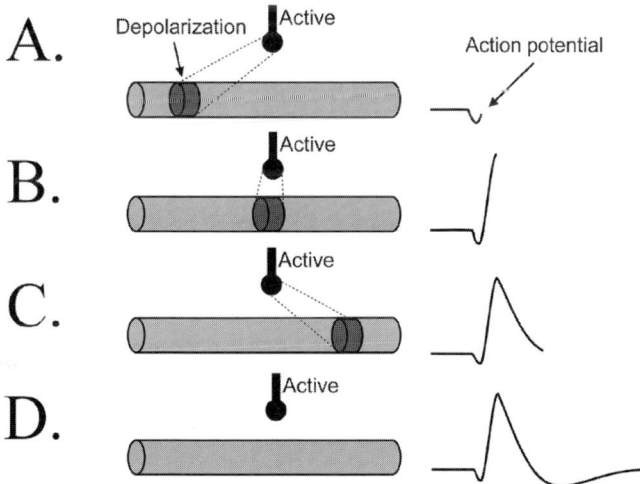

FIGURE 1.9 Far-field potential. **A.** Depolarization is approaching the active electrode. The approaching front produces a positive waveform. **B.** The depolarization travels beneath the electrode, so a negative potential is seen. **C.** The receding phase of repolarization produces a following positive phase. **D.** The potential has passed and the recorded potential returns to baseline.

overlying the nerve, and the reference (G2) is on the same limb but not overlying the nerve. As the wave of depolarization passes under G1, the inward current causes a prominent negative wave. As the wave passes, the potential returns to baseline.

Far-Field Potentials

Bipolar recording is similar to a unipolar recording except that G2 is over a distal segment of the nerve. The initial negative component is as previously described. As the depolarization passes under G2, however, a positive phase occurs because the amplifier is measuring the difference in potential between G1 and G2. Depolarization at G1 makes G1 negative relative to G2. Depolarization at G2 makes G2 negative relative to G1, but this means that G1 is positive relative to G2. Therefore, the recording is biphasic. The far-field potential is recorded with G1 at a distance from the current generator and G2 at a greater distance, typically in a different tissue compartment. The far-field potential is a stationary wave that is due to the moving front of depolarization in the axons. Since the far-field potential is recorded at a distance, it is not governed by exact electrode position. If the nerve volley comes close to the active or reference electrode, a component of near-field potential can be seen.

Most of the potentials recorded in clinical neurophysiology are far-field potentials. The notable exception is the compound action potentials of nerve conduction studies, which are largely near-field potentials.

2   Physics and Electronics

Atomic Structure

Atoms are composed of a central nucleus of protons and neutrons surrounded by electrons, which swirl around the nucleus. For most atoms, the number of negatively charged electrons is equal to the number of positively charged protons in the nucleus. The world as a whole is electrically neutral; however, some atoms have donated a "spare" electron to other atoms. Therefore, the atom that has donated the electron has a positive charge of +1, whereas the atom that has received the electron has a negative charge of −1. Figure 2.1 shows examples of neutral and charged atoms. The concept of donation and reception of electrons is based on the stability of full electron orbitals. Electron orbitals tend to be the most stable if they are filled with paired electrons. If one electron is present in the orbital, then the single electron is loosely held. Therefore, this electron is able to be donated to another atom. In this circumstance, there is then a net positive charge on the atom because of the loss of the electron. When this electron is incorporated into the partially filled orbital of another atom, this new atom now has a net negative charge, although this is an atomically stable situation since the orbitals are filled.

Conductors, Nonconductors, and Semiconductors

Conductors are materials that have partially filled orbitals that can accept electrons and have loosely held electrons that can be donated to nearby atoms, as shown in Figure 2.2. Conductors may be metals, some synthetic materials with conducting atoms, and ionic fluids such as biological tissues.

Nonconductors do not have partially filled electron orbitals or loosely held electrons. Therefore, even strong electric or magnetic fields cannot make electrons flow.

Semiconductors are nonconductors that have a small amount of conductor within the material. The introduction of conducting atoms in a controlled fashion within nonconducting materials is termed *doping* and is discussed further below.

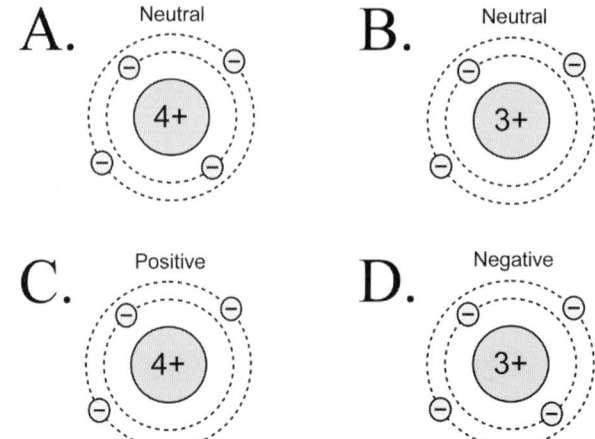

FIGURE 2.1 Diagram of atomic structure. The rings signify the outer orbitals and the center circle is the nucleus with the protons associated with the outer orbital shell. **A.** Atom has a full outer shell and is electrically neutral. **B.** Atom has a partially filled outer shell, and is electrically neutral. This atom can accept an electron to fill the outer shell, which would be atomically stable yet electrically charged. **C.** Atom from A has lost an electron and is now positively charged. **D.** Atom from B has gained an electron and is negatively charged. The filled orbital shell is atomically stable even though it is not electrically neutral.

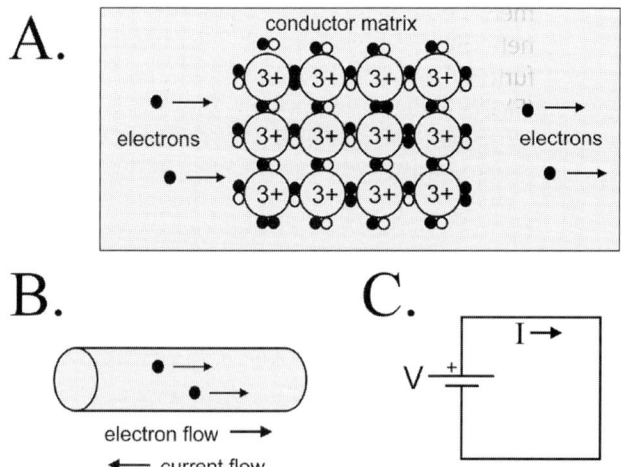

FIGURE 2.2 Diagram of a conductor. **A.** Atomic structure of a conductor. Adjacent atoms share electrons, shown as small black circles. The small white circles are "holes" that are unfilled portions of orbitals. **B.** Less enlarged view of a conductor, showing electron flow through the conductor. By convention, current flow is defined as the flow of positive charge, therefore, in the opposite direction to flow of electrons. **C.** Simple circuit diagram of a conductor. The symbol for a battery "V" has the positive terminal upward. Electrons flow from the negative terminal (bottom) through the conductor to the positive terminal. Current flow is opposite to the flow of electrons.

Fields

Fields are forces that have the ability to move electrons. There are two types: electric and magnetic.

Electric Fields

Electric fields are created by separation of charge. If electrons and their atoms are separated by a distance, there is an electric field between the two points of the charges. In fact, there is an electric field between every pair of charged particles. If the charges are same (e.g., both negative) the field repels the two particles. If the charges are opposite (e.g., positive–negative) then the field causes attraction between the particles.

Electric fields can be generated in a number of ways, including batteries, other power supplies, and biological currents. Electric fields cause electrons to move if connected by a conducting material.

Magnetic Fields

Magnetic fields can cause electrons to flow, especially if the magnetic field is changing. The magnetic field induces movement of electrons in a conducting material. Magnetic fields are generated in a variety of ways, including spin-orientation of atoms in a material and even electron movement itself. Movement of electrons in a conducting material causes a magnetic field that may in turn affect the flow of electrons. This is discussed further later. The electric and magnetic fields produce an *electromotive force* (EMF), which is just what it sounds like—a force that causes motion of electrons. The EMF has a vector that represents the direction and magnitude of the force.

Current Flow

Current flow is the movement of charge. Electrons are mobile, whereas nuclear protons are not. So in an electronic circuit, current is the flow of negatively charged electrons. Electrons flow from one atom to another, essentially jumping from one partially filled orbital to the next.

Electrons are constantly moving, but the movement is random. This is not considered to be current because there is no net flow of electrons in one direction, that is, there is no net vector of electron flow.

If an electric or magnetic field is applied to a conductor, current flows. The field causes electrons to flow in a unified direction so that there is a defined vector of electron flow. The unit quantity of charge could be the electron (−1) but other atomic particles have different valencies, including sodium (+1) and calcium (+2). The standard unit of charge is the *coulomb*. One coulomb is equivalent to 6.24×10^{18} positive or negative charges.

Current is movement of charge. The standard unit of current measurement is the *ampere* or *amp*. One amp is equal to one coulomb of charge passing through a point in a conductor each second, or

$$\text{Current} = \frac{\text{charge}}{\text{time}}$$

or

$$I = \frac{Q}{t}$$

We have been discussing flow of electrons, therefore negative charge, although conventional terminology considers flow of current to be flow of positive charge. While we can consider positive charge to flow through a conductor in one direction, we know that in reality, the current is flow of electrons in the opposite direction.

Circuit Theory

A circuit is a closed loop or series of loops composed of circuit elements. The connectors are conductors, usually wire or another conductor on a printed circuit board. The circuit elements can be:

- Resistor
- Capacitor
- Inductor
- Transistor
- Source of electromotive force

Most other circuit elements are components made from these elements.

Circuit Elements

Resistors

A resistor opposes the flow of electrons. Functionally, the resistor turns energy associated with the electrons into heat. Physically, the resistor dissipates energy imparted to the electrons by the electric field into heat. This is more than a semantic difference. Electrons leaving the negative pole of a battery are flowing down the potential gradient established by the chemical reaction in the battery, as shown in Figure 2.3. The electrons will flow quickly down the conductor, approaching the speed of light. However, the resistor dissipates some of the imparted energy into heat, so that there is less energy associated with the electrons. Therefore, fewer electrons make the journey. The amount of current is reduced. Resistance is measured in *ohms*, after the nineteenth-century German physicist Georg Ohm. If a voltmeter is placed across the terminals of the resistor, the recordable voltage difference reflects the energy dissipated by the resistor. This is the *voltage drop*. This is distinct from the *voltage source*, which would be the battery or power supply.

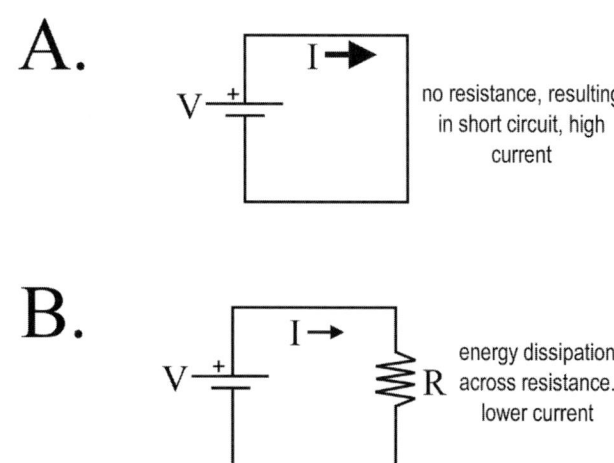

FIGURE 2.3 Theory of resistors. **A.** A conductor connects the terminals of a battery. Current will flow from the positive terminal to the negative terminal through the conductor. Electrons are flowing in the opposite direction. The magnitude of current will quickly destroy the battery. **B.** A resistor is inserted into the conductor circuit. This reduces the current flow by an amount that is proportional to the resistance of the resistor.

The voltage drop is finite when current is flowing. If no current is flowing, there is no voltage drop across the resistor.

Capacitors

Capacitors are thin plates of conductor separated by a nonconductor. Therefore, there is no direct conductive continuity between the terminals of the capacitor, as shown in Figure 2.4. Electrons cannot flow directly through a capacitor because of this separation of conductors; however, there is an apparent current, called a *capacitative current*. Electrons go from the negative terminal of the battery and onto one plate of the capacitor. The accumulation of electrons on the plate produces an electric field that repels electrons on the nearby plate of the capacitor. Electrons flow off this plate, through the conducting wire and into the positive terminal of the power supply. The electrons entering and leaving the capacitor are not the same electrons, but they seem so. Therefore, this is a capacitative current.

As capacitative current flows, a charge is built up across the capacitor. This charge opposes the charge of the power supply. The charge across the capacitor opposes the charge of the power supply, so there is less net force pushing electrons onto the capacitor. When the voltage across the capacitor is exactly equal to the voltage of the power supply, movement of electrons ceases. The capacitor is fully charged. When the power supply is turned off, the only voltage providing EMF to the circuit is that of the capacitor. Electrons flow off the plate, through the conductor, to the opposite plate of the capacitor, thereby discharging the capacitor. When the voltage across the capacitor is discharged, current ceases.

Capacitance is a measure of the capacity to store charge. Measurement is in *farads*, after Michael Faraday, the nineteenth-century English physicist.

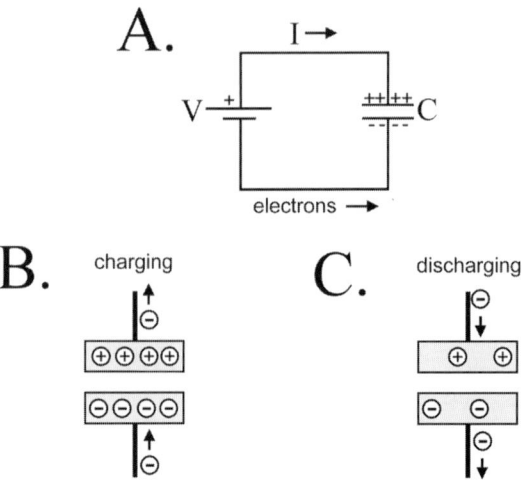

FIGURE 2.4 Theory of capacitors. **A.** Simple circuit of a battery charging a capacitor. When the battery is switched on, electrons flow from the negative terminal and onto the lower plate of the capacitor. Electrons leave the upper plate and travel through the conductor to the positive terminal of the battery. The departure of the electrons from the upper plate produces positively charged holes. **B.** Close-up of the capacitor during the charging phase. Electrons accumulate on the lower plate and depart the upper plate. **C.** When the battery is switched off while the terminals of the capacitor are connected, electrons flow off the lower plate and onto the upper plate. The motivation for the electron movement is the charge separation across the plates of the capacitor.

Higher capacitance means that more charge is built up on the capacitor for a given voltage. The current that flows onto the capacitor is proportional to the capacitance and the change in voltage. Expressed mathematically:

$$I = C \frac{dV}{dt}$$

where C is capacitance, dV is change in voltage, and dt is change in time. Therefore, the current that can flow through a capacitor is proportional to the capacity or capacitance of the capacitor, and proportional to the rate of change in voltage. This latter relationship is because the current is greatest soon after a change in voltage.

Inductors

Induction is the effect of magnetic fields on charged particles. When electrons pass through a wire, there is a magnetic field that is oriented radially around the wire. The orientation of the magnetic field is governed by the *right-hand rule*, a law that is often remembered from physics, although the application is forgotten. With the fingers of the right hand closed and the thumb extended, current flow along the axis of the thumb produces a magnetic field that is oriented as the curled closed fingers of the hand. This is the right-hand rule for conventional current.

The magnetic field can influence the movement of charge within the wire itself and even influence the movement of charge in adjacent conductors that are not part of the circuit.

Coiling the wire results in alignment of multiple turns of the wire. When electrons flow in the wire, the magnetic fields sum in a complex way, so that a huge magnetic field can be generated. The distribution of the magnetic fields is shown in Figure 2.5. The theory of inductor function is diagrammed in Figure 2.6.

An *inductor* is a circuit element that consists of a coil of wire. There may be a metal core in the center of the coil. An inductor stores energy as a magnetic field. This is similar to a capacitor, which stores energy by separation of charge, although the effects on the circuits are very different. As current flows through the wire, some of the energy of the electrons is used to generate the magnetic field. When the power supply is turned off, the magnetic field collapses, imparting some EMF to the electrons. This causes current to flow for a time after the power supply has been turned off.

Behavior of an inductor differs from that of a capacitor in the following ways:

- Opposition to flow of current decreases when an inductor is charged, while it increases with charging of capacitor.
- Terminal current flow (after the power supply is turned off) is forward for an inductor but the reverse for a capacitor.
- There is a limit to the charge that can pass through a capacitor but this is not so with an inductor.

Semiconductors

Semiconductors are so called because they conduct better than nonconductors but less well than conductors. In general, semiconductors are made from nonconducting material that has a small amount of conducting material mixed in. This mixing process is termed *doping,* and when the nonconductor has the conducting material in it, it is *doped*.

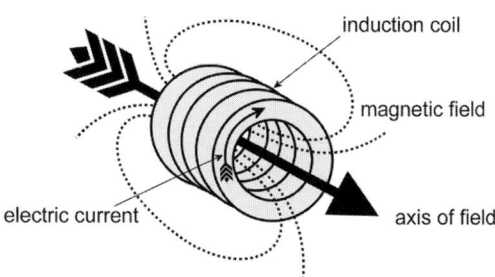

FIGURE 2.5 Magnetic fields created by conductors and inductors. Current flow through a conductor produces a magnetic field surrounding the conductor. When the conducting wire is coiled, the induced magnetic fields sum to produce a radially oriented field. The field has a geometry, strength, and axis, indicated by the arrow.

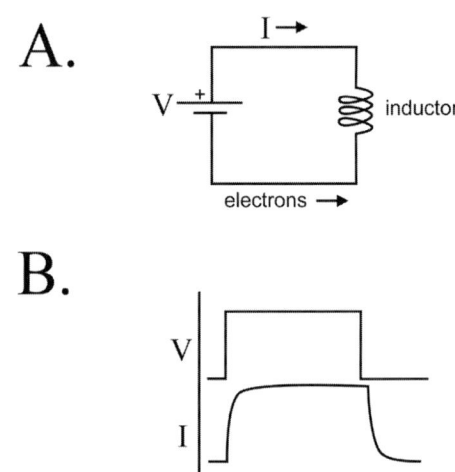

FIGURE 2.6   Theory of inductors. **A.** An inductor is part of a circuit with a battery. As current flows through the inductor, a magnetic field develops around the inductor. **B.** Current through the circuit as a function of voltage. As the voltage is changed, the current lags during charging of the inductor. When the voltage is switched off and the terminals of the inductor are still connected, current flows through the circuit by induction, that is, collapse of the magnetic field produces movement of electrons through the conductor for a short time.

The most common nonconductor used for semiconductors is silicon. This is the basic element of glass and sand. Silicon has a full outer electron orbital where there are no unpaired electrons. Therefore, silicon is unwilling to donate an electron for current flow, and has no space to accept an electron for current flow. Gallium and arsenic are conducting atoms that are used to dope silicon for the manufacture of semiconductors, and there are several others, as shown in Figure 2.7.

Arsenic has a single unpaired electron that is loosely held. This electron can easily be donated to a nearby partly filled orbital, leaving the arsenic atom with a charge of +1, due to the loss of electron. Since the loosely held electron is negatively charged, this semiconductor is *negatively doped* or *N-doped*.

Gallium has an empty orbital in an otherwise filled outer shell of electrons. Gallium can easily accept an electron to fill its outer orbitals, giving it a charge of –1 with the extra electron. Since the extra space can accept a negatively charged electron, this semiconductor is *positively doped* or *P-doped*.

Current can easily flow through a semiconductor, though less well than through a conductor. When only one type of doped material is used, current can flow in both directions. The magic begins when N-doped and P-doped materials are placed together.

Diodes

Diodes are composed of N-doped and P-doped material placed together, as shown in Figure 2.8. One terminal of the diode is connected to the N-doped material and the other terminal is connected to the P-doped material.

The region of interface between the N-doped and P-doped materials is of special interest. Loosely held electrons from the N-doped side move across

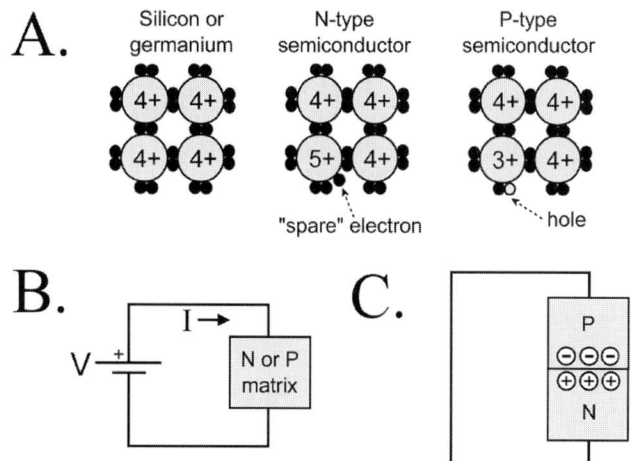

FIGURE 2.7 Semiconductor theory. **A.** Atomic structure of semiconductor material. Silicon (left) is a nonconductor. Doping of this material with a conducting element results in either available electrons for flow (center, N-type), or available empty electron orbitals that can temporarily host a flowing electron ("hole" or P-type). **B.** A piece of semiconductor material as part of a circuit can conduct electric current, though not as well as a conductor. **C.** Two semiconductor pieces are placed together without a battery. Some of the spare electrons of the N-type semiconductor move to occupy partially filled orbitals of the P-type semiconductor. This is analogous to the diffusion potential of biological membranes.

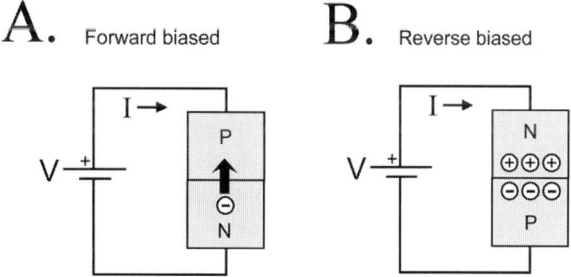

FIGURE 2.8 Diode theory. **A.** A battery is connected to the N–P semiconductor complex. In this diagram, current can flow, since the junction potential between the two materials is negated by the applied voltage. **B.** The position of the semiconductor materials is reversed so that the applied voltage serves to augment the junction potential. Current cannot flow. Since the N-P semiconductor junction allows for current flow in only one direction, this is a diode.

the junction to the P-doped side thereby occupying the empty spots in the electron orbitals. This situation is not electrically neutral but is atomically quite stable; the N-doped material has a positive charge because of the electrons being donated. The P-doped material has a negative charge because of the acceptance of electrons. The migration of electrons continues until a charge is built up across this junction that is sufficient to halt the further migration of electrons. This electrical equilibrium is similar to the diffusion potential of excitable tissues, where potassium passes through the membrane down its chemical gradient until sufficient electric charge is built up across the membrane to stop further migration.

A diode takes advantage of the junction effect between semiconductors to allow current movement in only one direction. If a battery is connected to the diode with its positive terminal attached to the P-side and the negative terminal attached to the N-side, then the diffusion potential at the NP junction is abolished and current can easily pass through the diode. Since current is passed, this diode is *forward biased*.

On the other hand, if the positive terminal is attached to the N-doped side and the negative terminal is attached to the P-doped side, the junction potential is enhanced. All of the loosely held electrons near the junction have passed from the N-side to the P-side. Virtually all of the empty electron orbitals on the P-side at the junction are filled with those loosely held electrons. Therefore, there are no more loosely held electrons nor available holes (empty electron orbitals) for electrons to pass through; current does not flow. Since current cannot flow in the desired direction, this diode is *reverse biased*.

This is the essential function of a diode; current can pass in only one direction. In our modern electrical circuitry, diodes are most important for turning alternating current (AC) into direct current (DC), since most solid-state electronics uses DC rather than AC to function. The AC input is directed into two pathways. One pathway uses a diode to allow positive voltage to drive current through the diode but blocks the negative voltage. The other pathway has the diode reversed so that negative voltage drives current through the diode but positive current is blocked. Then, the voltage from this pathway is reversed, negative becoming positive. The two pathways then rejoin, and there is an almost continuous voltage of a positive potential. This device is called a *rectifier*, and is the most important use of diodes.

Transistors

The term *transistor* is derived from the words *transfer* and *resistor*, since the transistor controls the transference of energy across a resistance. Transistor theory is summarized in Figure 2.9. Transistors have replaced the thermionic tubes used in older equipment. While we used to describe transistors in comparison with tube function, the reader's frame of reference has changed such that this discussion is of little value. A better analogy might be an automobile.

If one crosses town using a bike, the speed of the bike is dependent on the force applied by the feet to the pedals. Although there is some mechanical advantage to the bike, including gear ratios and the ability to coast, in general, all of the energy used to move the bike and body is supplied by the feet. If the same journey is made using an automobile, the power to move across town is provided by the engine, but the speed of the engine is controlled by a foot on the accelerator pedal. The speed of the car is proportional to the power supplied by the engine, which in turn is proportional to the pressure of the foot on the pedal. The power of the foot is essential for the journey, but the foot muscles had help, that is, the engine.

This analogy defines the difference between a *passive circuit* and an *active circuit*. A passive circuit has all of the energy to the system supplied by the applied signal voltage (as for the bike). The active circuit uses an internal energy supply so that the output of the circuit can be greater than the input signal voltage (as for the car).

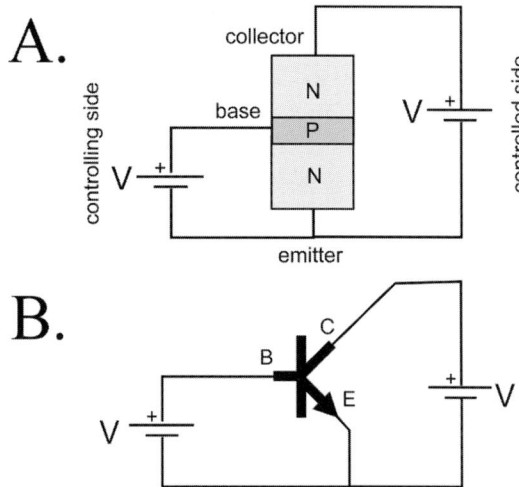

FIGURE 2.9 Transistor theory. **A.** A transistor is constructed from three layers of semiconductor material. Voltage applied on the left side of the circuit controls the conductance across the entire transistor. Therefore, a small voltage on the controlling side governs current flowing on the controlled side. **B.** Circuit diagram of the transistor shown above.

What does this have to do with transistors? The transistor is the essential element in the car analogy. The applied signal voltage is the foot on the accelerator pedal. The engine is the device power supply. The current through the output impedance is the power delivered to the wheels. The output voltage is directly proportional to the signal voltage, though greatly amplified.

The amplification for most transistors is 9×, but for simplicity of math, we will consider it to be 10×. If three amplifiers are placed in series, the amplification is 1000×. For most display systems, the amplification may need to be in the millions.

Circuit Laws

There are three basic laws governing circuits that should be discussed. The laws are not used in daily electrophysiologic practice, but are important for understanding the electrical behavior of circuits. The laws are:

- Ohm's law
- Kirchhoff's current law
- Kirchhoff's voltage law

Ohm's Law

Ohm's law describes the relationship between voltage (V), resistance (R), and current (I) for a resistive circuit. Figure 2.10 shows a simple resistive circuit

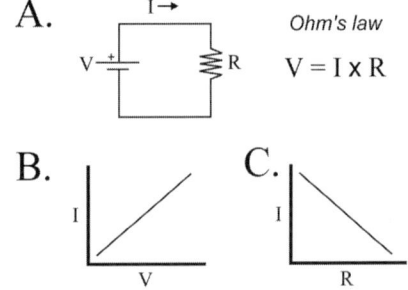

FIGURE 2.10 Ohm's law. **A.** Simple resistive circuit with a battery in series with a resistor. A current flows through the resistor that is proportional to the voltage and inversely proportional to the resistance of the resistor. **B.** Direct relationship between current and applied voltage. **C.** Inverse relationship between current and resistance.

with a power supply connected to a resistor. For this circuit, the supply voltage is equal to the current multiplied by the resistance:

$$V = I \times R$$

where voltage is measured in volts, current in amps, and resistance in ohms.

Ohm's law: The applied voltage is equal to the current multiplied by the resistance of the circuit.

The implications of Ohm's law are evident from permutations of the law:

- Increasing voltage produces an increase in current.
- Increasing resistance produces a decrease in current.

The formula may not seem intuitively obvious, but the relationship becomes clearer when these permutations of the formula are considered. First, one can expect that an increase in voltage would increase the current passing through a fixed resistance. Also, it is easy to imagine that increasing the resistance would decrease the current flow produced by a fixed voltage.

The best analogy to this circuit is a pail of water. The pail is partly filled with water and there is a small hole in the bottom of the pail. The deeper the water, the more pressure there is on the water to flow out of the hole and the more water runs out. Likewise, making the hole smaller decreases the amount of water that can exit the pail. In this analogy, the water passing through the bottom of the pail is the current, the depth of the water represents the voltage, and size of the hole represents the resistance—with a smaller hole having higher resistance to flow.

Ohm's law may seem simple, but it is fundamental to all of the other laws governing circuit function.

Kirchhoff's Current Law

A node is a connector point, essentially a soldered or etched connection between two or more conductors. A node has no ability to change or store energy, no resistance or capacitance, and insignificant magnetic properties. Therefore, all of the electrons flowing into a node must flow out through one of the conductors with negligible loss of EMF. Kirchhoff's current law is a statement of this physical property.

For any node, the sum of the currents flowing into a node is equal to the current flowing out of a node. Since current flowing out of a node can be considered to be of opposite sign to current that is flowing into a node, then, the statement becomes:

> Kirchhoff's current law: The sum of currents flowing into and out of a node is zero,

or

$$\sum I = 0$$

Kirchhoff's Voltage Law

Kirchhoff's voltage law is not as intuitively obvious as the current law. This law states that for any circuit loop, the sum of the voltage sources equals the sum of the voltage drops, or:

> Kirchhoff's voltage law: For any circuit loop, the sum of the voltage sources and voltage drops is equal to zero.

- A *circuit loop* is any combination of circuit elements connected by a conductor. In Figure 2.11, this law applies to any combination of loops

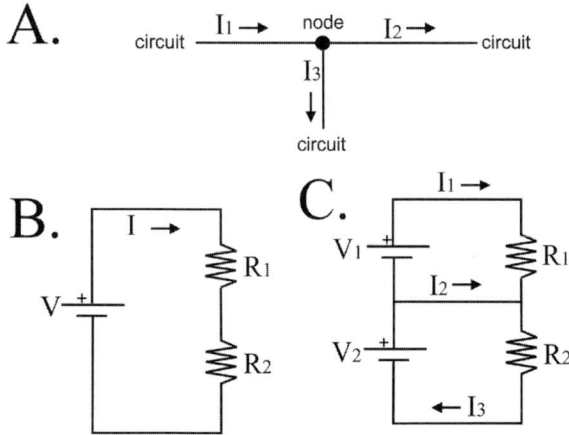

FIGURE 2.11 Kirchhoff's laws. **A.** Kirchhoff's current law—the sum of the currents flowing into and out of a node is zero. The node cannot store or modify energy. **B.** Kirchhoff's voltage law—the sum of the voltage sources and drops in a circuit loop is zero. **C.** Kirchhoff's voltage law applies to all circuit loops, including each of the smaller loops in this diagram and the large loop encompassing both batteries and both resistors.

including the top loop, the bottom loop, and the large loop encompassing both power supplies and both resistors, without the middle conductor.
- *Voltage sources* are batteries or power supplies.
- *Voltage drops* are circuit elements that reduce EMF to the electrons, and for this example are only resistors. A voltage drop is measured by placing a voltmeter across a resistor, but can be calculated from Ohm's law by multiplying the current times the resistance of the resistor ($V = I \times R$).

Consider one of the circuit loops in Figure 2.11. If the power supply voltage is increased, the current increases since the resistance is fixed, from Ohm's law. Increased current results in increased voltage drop across the resistor that is equal to the increased voltage of the power supply.

Circuit Properties

Ohm's and Kirchhoff's laws can be used to derive other circuit laws, but the mathematical derivations are unimportant for clinical use. A basic understanding of the principles is essential.

Resistors

Basic resistor theory was presented previously. Resistor theory continues with discussion of resistors in series, resistors in parallel, and finally, the RC circuit.

Series Resistors

Resistors are often placed in series, as shown in Figure 2.12. The total resistance of two or more resistors in series is equal to the sum of the individual resistances:

$$R_{total} = R_1 + R_2$$

or

$$R_{total} = \Sigma R_i$$

This would be expected from Kirchhoff's voltage law, since the power supplied to the circuit (V_s) equals the sum of the voltage drops across the resistors (V_{R1} and V_{R2}):

$$V_s = V_{R1} + V_{R2}$$

Since $V = I \times R$ for each resistor,

$$I \times R_{total} = (I \times R_1) + (I \times R_2)$$

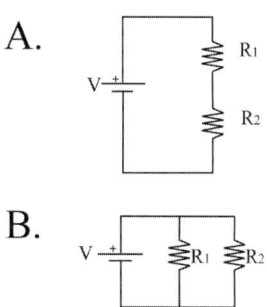

FIGURE 2.12 Additive resistances. **A.** Series resistors---the total resistance of two resistors in series is equal to the sum of the individual resistances. **B.** Parallel resistors---the reciprocal of the total resistance of two resistors in parallel is equal to the sum of the reciprocals of the individual resistances.

I is the same for the power supply and both resistors since there is only one pathway for current to flow. Therefore, *I* drops out, leaving

$$R_{total} = R_1 + R_2$$

or

$$R_{total} = \Sigma R_i$$

where R_i is the resistance of each resistor.

Parallel Resistors

Parallel resistances are conceptually somewhat more complex. First, we will discuss the concept, then the mathematics.

Figure 2-12 shows the simplest circuit involving two resistors in parallel. The total resistance of the two resistors is *less* than each of the individual resistances. An analogy would be the pail half-full of water previously discussed. If one tiny hole is punched into the bottom of the pail, the water will run out, albeit slowly through this high resistance. If a second hole of the same size is punched through the bottom of the pail, each hole has high resistance, but there are now two avenues for flow of water, so the total resistance is less. The same applies to our circuit. There are two avenues for electrons to flow through resistors to the positive terminal, so the total resistance is less than the resistance of each resistor. How much less?

To simplify the math, let's consider conductance. Conductance (*G*) is the reciprocal of resistance. A material of lower resistance conducts better. Therefore

$$G = 1/R \text{ for each resistor}$$

Back to the pail analogy: each hole has a conductance that describes the flow of water for a specific depth of the column of water (analogous to voltage). The total conductance of the pail is the sum of the conductances of the individual holes:

$$G_{total} = G_1 + G_2$$

or

$$G_{total} = \Sigma G_i$$

where G_i is the conductance of the individual holes, or resistors. Since $G = 1/R$, this formula becomes

$$\frac{1}{R_{total}} = \frac{1}{R_1} + \frac{1}{R_2}$$

or

$$\frac{1}{R_{total}} = \Sigma \frac{1}{R_i}$$

Therefore, the reciprocal of the total resistance is equal to the sum of the reciprocals of the individual resistances. This is difficult to conceptualize unless we think of conductance.

Capacitors

Capacitors can be placed in series and parallel, just as resistors, but because of the electrical properties, the formulae are reversed! For simplicity, parallel capacitors will be discussed first.

Parallel Capacitors

Figure 2-13 shows the simplest circuit diagram of capacitors in parallel. This serves essentially to increase the size of the plates of the capacitor. There is still no direct route for electrons to go from the negative terminal to the positive terminal. The total capacitance of the system is equal to the sum of the individual capacitances in parallel:

$$C_{total} = C_1 + C_2$$

or

$$C_{total} = \Sigma C_i$$

Series Capacitors

Capacitors in series have a complex interaction. The charge developed from charging of one capacitor affects the time to charge the second capacitor. In practice, the effect is affects both capacitors reciprocally. The math is more complex than for resistances, but the result is similar. For two capacitors in series, the reciprocal of the total capacitance is equal to the sum of the

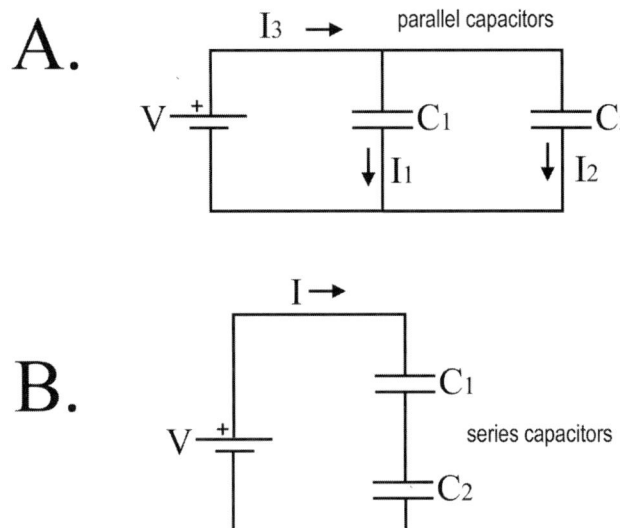

FIGURE 2.13 Additive capacitances. **A.** Parallel capacitors—the total capacitance of two capacitors in parallel is equal to the sum of the individual capacitances. **B.** Series capacitors—the reciprocal of the total capacitance is equal to the sum of the reciprocals of the individual capacitances.

reciprocals of the individual capacitances:

$$\frac{1}{C_{total}} = \frac{1}{C_1} + \frac{1}{C_2}$$

or

$$\frac{1}{C_{total}} = \Sigma \frac{1}{C_i}$$

Filters

All biological signals can be broken down into fundamental frequencies, with each frequency having its own intensity. Display of the intensities of all frequencies is a *power spectrum*. In clinical neurophysiology, we usually are interested in signals of a particular frequency range, termed the *bandwidth*. The bandwidth differs for different studies. The bandwidth is determined by filters—devices that alter the frequency composition of the signal. There are three basic types of filter:

- Low-frequency filter (LFF)
- High-frequency filter (HFF)
- 60-Hz (notch) filter

The LFF is sometimes called the high-pass filter, but this old terminology should be discarded. The LFF filters out the lower frequencies. The HFF (also formerly called the low-pass filter) filters out high frequencies. Particular settings of the filters determine the exact frequencies that are affected.

The *60-Hz filter* uses tandem LFF and HFF to attenuate the 60-Hz powerline frequency. Unfortunately, the filter is not perfect, and frequencies close to this are attenuated, as well. Therefore, the appearance of the biological signal may be altered by use of the 60-Hz filter. This is especially true for EEG.

Resistor–Capacitor Circuits

The simplest filter is not used in most neurodiagnostic equipment but discussion is helpful for understanding the concept. A resistor and a capacitor in series forms the resistor–capacitor or *RC circuit*. This is shown in Figure 2.14.

The resistor (R) and capacitor (C) are in series with a power source that is a signal voltage (V_S). The effects of a sudden change in voltage on current are shown in the figure. Immediately after the voltage is turned on, current (I) begins to flow between the terminals of the resistor. This current causes a build-up of charge on the capacitor as capacitative current flows. The charge on the capacitor opposes flow of further current, ultimately stopping current flow when the voltage across the capacitor is equal and opposite to the signal voltage.

When the signal voltage is reduced to zero, the capacitor is not the only source of EMF in the circuit, so it discharges; electrons flow through the circuit again, but in the reverse direction to the initial current flow when the capacitor was charging.

This means that when there is a sudden change in voltage, the voltage is high across the resistor and low across the capacitor. At steady state, the voltage is maximal across the capacitor and zero across the resistor. Essentially this is functioning as a filter. If you consider this signal voltage and look at the voltage across the resistor and capacitor separately, the resistor has filtered out the low-frequency steady-state potential but shown the higher frequency voltage change. In contrast, the capacitor has filtered out the high-frequency step in voltage, but shown the plateau potential.

The *time constant* (TC) represents the amount of time required for the capacitor of an RC circuit to charge or discharge to $1/e$ of the plateau, where

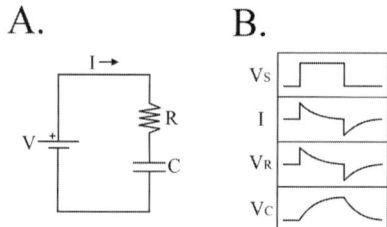

FIGURE 2.14   *Resistor–capacitor circuit.* **A.** *A resistor and capacitor are in series with a battery.* **B.** *When the battery is switched on, there is an immediate current that produces a voltage drop across the resistor (V_R). As current flows, the capacitor is charged (V_C); this potential opposes the subsequent flow of current.*

e is the natural base, approximately 2.7. This value is independent of applied voltage, as long as the voltage is constant after a change.

Active, Passive, and Digital Filters

The RC circuit is the simplest filter. Since no exogenous energy is required to make the filter work, this is a *passive filter*. Most analog filters are *active filters*, using power supplies to run circuits that alter the frequency composition of the signs. *Digital filters* change the frequency response of the signal by performing calculations on the data. Digital filters are a major part of most neurophysiologic equipment.

Frequency Response

All filters result in a particular frequency response. The filters do not cut off all frequencies outside of the set range. There is a fall-off in response so that frequencies far below the LFF setting are reduced much more than frequencies a little below the LFF setting. The same holds for frequencies above the setting of the HFF. Figure 2.15 shows frequency response for a number of filter settings. Most equipment comes with default settings for filters that are reasonable. However, these default values should be checked, since there are occasionally programming errors at the factory. We have already experienced this on one of our machines; the default settings for median sensory nerve conduction studies (NCS) were far different from that for any other sensory NCS, and were clearly wrong!

It is instructive to record EEG and EMG and play with the filter settings. This is best done using digital EEG where the same epoch of activity, slow, fast, and spike, can be replayed with different recording parameters.

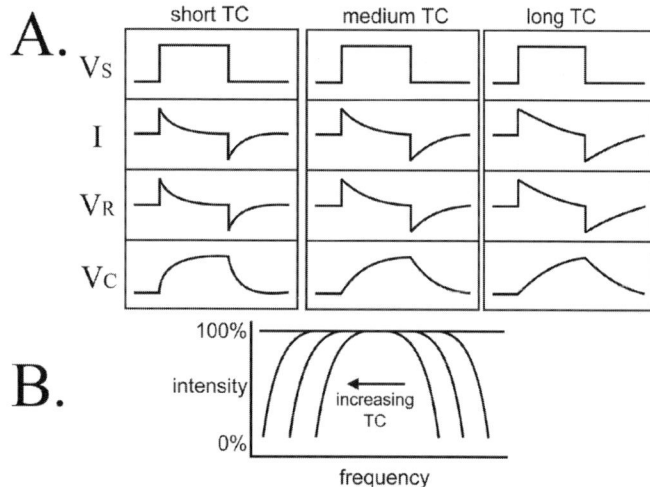

FIGURE 2.15 *Frequency response of filters.* **A.** *Effect of time constant on current (I), voltage across the resistor (V_R), and voltage across the capacitor (V_C) of an RC circuit.* **B.** *Frequency-response curves with the RC circuit with different time constant. Increasing time constant pushes the curves toward the lower frequencies. The decline in response to low frequencies is V_R. The decline in response to high frequencies is V_C.*

Of course, remember to return the settings to their correct values before the next diagnostic study!

Digital Signal Analysis

Analog-to-Digital Conversion

Many of us were trained in the time of dinosaurs when we had to pay close attention to the details of analog-to-digital conversion (ADC). Conversion cards were expensive, of low performance, and needed software that we often wrote ourselves. We did not have ADC hardware on inexpensive boards that could turn a laptop into a powerful neurodiagnostic machine. It is interesting to note that every computer has an ADC device—it is called the sound card, although the performance is nothing like the ADC chips available in modern equipment.

Details of the function of ADC devices are presented on the CD, since the function is unimportant for routine interpretation. Suffice it to say that the ADC device samples the signal voltage at predetermined times. This results in an array of digital values, where each point has three values—channel ID, time, and voltage. These data can then be manipulated with simple calculations.

Calculations

The digital data can be manipulated for a variety of purposes. Some of these are:

- Averaging
- Spike detection
- Digital filtering

Digital filtering is the most common application. Calculations performed on the digital data can remove unwanted frequencies. This can change the appearance of the recording, so the commonly held impression that frequency and phase distortion does not occur with digital filtering is unfortunately not true.

Averaging is used especially for evoked potential studies, although sensory NCS also employ averaging when there is difficulty in eliciting a good measurable response.

Spike detection is used predominantly for patients with suspected epilepsy. Calculations performed on the digital EEG data can tell whether there has been sharp activity superimposed on an otherwise normal background.

Additional detail is presented on the CD.

Electrodes and the Patient–Equipment Interface

Electrodes are merely connectors, allowing the patient and the neurophysiologic equipment to be part of the same circuit. Unfortunately, Kirchhoff's current law does not apply to the node that is an electrode, because the connection between the electrical and biological conductors is complex. There is some ability of the junctions to store and modify energy.

Disk Electrodes with Gel

Disk electrodes are used mainly for EEG and EP. Their interface is more complex than it appears. The gel acts as a malleable extension of the electrode, so that movement of the electrode leads is less likely to produce artifacts. The gel maximizes skin contact and allows for a low-resistance recording through the skin.

Electrodes may be reversible or nonreversible. This is detailed on the CD. Reversible electrodes include the commonly used silver chloride electrodes. Reversible means that the chemical reaction for electron transfer is bidirectional. This is the best for routine recording. Nonreversible electrodes have difficulty with electron flow in one direction, essentially having the junction function as a diode plus a capacitor. Current flows in one direction but not the other. When current tries to flow in the opposite direction, a potential is built up across the electrode junction and current flow is further impeded.

Needle Electrodes

Needle electrodes are used almost exclusively for EMG. While needle electrodes have been used for EEG in the past, this application has been largely abandoned. For EMG, the needle is inserted into the muscle. The voltage changes of nearby fibers cause a very small amount of current to flow through a coaxial electrode and through its leads to the amplifier before completing the circuit through the reference electrode and tissue. Although the electrode impedance is fairly small, the amplifier input impedance is large, so the actual amount of current flowing into the amplifier is small.

Patient–Equipment interface

The electrode–amplifier interface is a key part of the patient–equipment interface. Figure 2.16 shows the interface as a circuit representation. V_s is the signal voltage and is generated by the neurons. The active electrode is the top lead of V_s, while the reference electrode is the bottom lead. R_e is the resistance of the electrode attached to the patient. This is not a resistor but rather is a product of the electrical properties of the electrode attached to the skin. R_{in} is the input resistance of the amplifier. Again, this is not a fixed resistor, but rather an effect of the electrical circuitry.

The voltage passed to the power amplifiers and display is the voltage drop seen across the input resistance of the amplifier, R_{in}. Therefore, anything that alters voltage across this resistance alters the measurement, distorting the recording. This can occur in the following circumstances:

- High electrode resistance (R_e)
- Significant electrode stray capacitance (C)

The derivation for the formula is on the CD, but the following relationship can be shown:

$$\frac{V_{in}}{V_s} = \frac{R_{in}}{R_e + R_{in}}$$

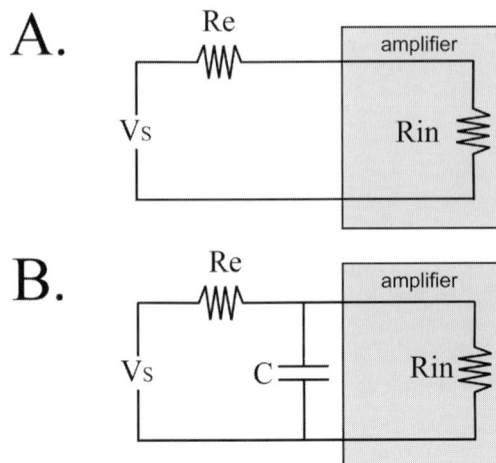

FIGURE 2.16 *Electrode–amplifier interface.* **A.** *Diagram of the electrode–amplifier interface. V_s is the signal voltage from the body, R_e is the electrode resistance, and R_{in} is the input resistance of the amplifier. The signal voltage is equal to the sum of voltages across the electrode (R_e) and the input resistance of the amplifier (R_{in}).* **B.** *Same diagram as in A, but a small capacitance (C) is inserted into the circuit. This capacitance is created by proximity of the electrode leads and electrodes.*

This means that the ratio of the voltage seen by the amplifier to the voltage generated by the signal source is dependent on the ratio of the input resistance to total resistance of the circuit. Therefore, high electrode resistance causes a higher voltage drop across the electrode, thereby reducing available voltage to the amplifier.

The second diagram in Figure 2.16 shows a capacitor (C) that is not a soldered capacitor, but rather a capacitance that is created by the proximity of the leads to the patient. With the electrode resistance plus this capacitance, this circuit acts like a filter, an RC circuit. Therefore, if the resistance and/or capacitance is significant, the appearance of the signal voltage may change dramatically.

In these discussions we have been discussing electrode and amplifier resistance. The better term would be impedance, since impedance is frequency-dependent resistance. The math is much more complicated for impedances, so for conceptual purposes we discuss resistances. Hereafter, we will use the more correct term impedance.

Amplifiers

Amplifiers use transistors, which were discussed previously. The concept is that a controlling signal voltage is increased in size by the amplifier. Ganging of multiple amplifiers is needed in order to produce the voltage needed to drive the display devices or the analog-to-digital converters of the computers—these devices work with fractions of a volt rather than millivolts.

The key to electrophysiologic equipment is the differential amplifier. This is shown in Figure 2.17. The major advantage for our work is the ability to

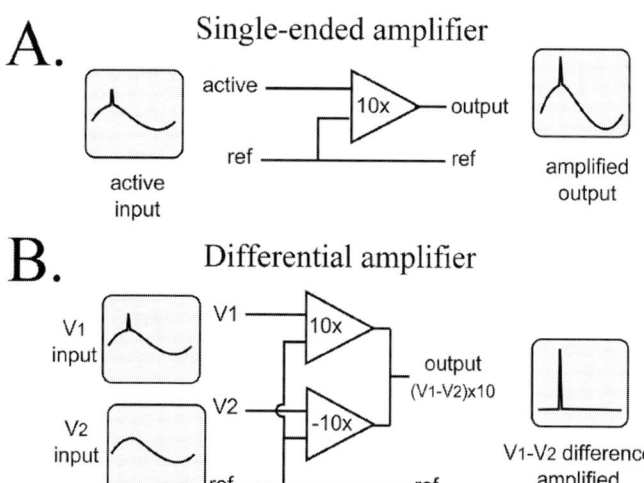

FIGURE 2.17 Differential amplifier. **A.** A single-ended amplifier takes the biological signal shown on the left and amplifies it to the amplified output shown on the right. The amplifier cannot distinguish between the signal from the active electrode and the signal from the reference electrode. **B.** A differential amplifier displays the magnified output of the difference between two inputs. Assuming that the unwanted signal is more likely to be common to the two inputs than the signal of interest, the amplified output subtracts the common signal or common mode.

eliminate much noise that is common to the two inputs to the amplifier. This is termed *common-mode rejection*. Since the driver amplifier sees the amplified difference between the inputs, the signal voltage is preferentially passed to the driver amplifier. Noise is likely to be in common to the two inputs, therefore, it is rejected from the output. The ability of a differential amplifier to reject common signal is described by the *common-mode rejection ratio*:

$$\text{CMRR} = \frac{\text{Common signal voltage}}{\text{Nonamplified output voltage}}$$

For most modern amplifiers, this ratio is at least 10,000 to 1.

Unequal electrode impedances can change the voltage seen by the input impedances of the amplifiers, as described above. Therefore, the ability of the differential amplifier to eliminate common mode is degraded by unequal electrode impedances.

Physics and Electronics

Part II Electroencephalography

3   Electroencephalography Basics

Physiology of Electroencephalography

Electroencephalographic (EEG) potentials are generated in the cerebral cortex, then conducted through the skull and scalp. Because of the electrical properties of the inactive tissues, potentials are of small amplitude and represent the summed activity of many neurons. Some of the recorded activity is generated by action potentials but most is generated by excitatory and inhibitory postsynaptic potentials (EPSPs and IPSPs). Basic principles of action potentials and synaptic transmission are presented in Chapter 1.

Generation of Electroencephalographic Rhythms

Scalp electrodes detect charge movement only in the most superficial regions of the cerebral cortex. Electrical activity in the deep nuclei produces surface potentials of too low amplitude to be detected reliably. These potentials are overshadowed by the cortical activity.

Cortical Potentials

Most of the cortical efferents have large cell bodies oriented perpendicularly to the cortical surface (Figure 3.1). While the cortex has a convoluted surface, the scalp EEG electrodes see best the electrical activity arising from the regions that are relatively parallel to the scalp. Inhibitory and excitatory inputs to these large efferent neurons produce substantial current that sums to become the scalp EEG. The dendritic arborization is in the superficial layers of the cortex, while the soma and axon hillock are in the deeper layers. This creates a vertical columnar organization of the cortex.

Activation of thalamocortical afferents results in EPSPs and IPSPs in the interneurons and efferent neurons. Depolarization of the dendrites is conducted along the cell membrane to the axon hillock, where an efferent action potential is generated. Efferent action potentials project to nuclei in the subcortex, brainstem, spinal cord, and other cortical regions. The influx of positive ions into the efferent neuron results in a negative extracellular field potential. Electrotonic depolarization of the soma and axon hillock results in a positive field potential. The vertical columnar organization of the

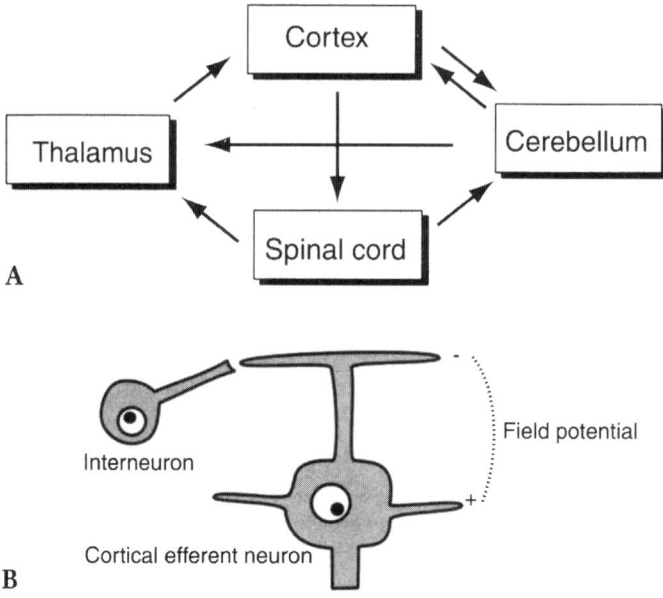

FIGURE 3.1 *Organization of the cerebral cortex.* **A.** *Some of the connections of the cortical and subcortical tissues are shown here. Projections from one section to another can produce potentials recorded during EEG and EPs.* **B.** *Columnar organization of the cerebral cortex, with the large cortical efferents oriented so that the dendritic arborization is near the surface.*

cortex makes the negative field superficial to the positive field, creating a *dipole*. While the term dipole is used to describe the positive–negative vectors of spikes and sharp waves, virtually all electrophysiologic activity can be represented as a positive–negative dipole, though not usually in a straight line.

When discussing EEG activity, whether normal or abnormal, we discuss rhythms. Neuronal electrical activity has an inherent rhythmicity, and the rhythms can be generated at several levels. Rhythmicity can be local to a neuron, regional involving neighboring neuronal circuits, or remote involving neuronal circuits that span brain structures. For example, single cells from the thalamus may demonstrate membrane potential oscillation in culture. Hippocampal slices develop rhythmic cellular discharge that is dependent on neuronal interaction. Many rhythms recorded from scalp electrodes are produced by synchronous rhythmicity involving thalamocortical circuits. These three examples demonstrate that rhythmicity is a fundamental property of many neurons and neuronal circuits that is used for generation of physiological activity such as motor system activation as well as pathological activity such as convulsions.

Scalp Potentials

Scalp electrodes are not able to detect all of the charge movement occurring at the cortical surface. One estimate suggests that 6 cm² of cortical surface area must be synchronously activated for a potential to be recorded at the scalp. Potentials are volume-conducted through the meninges, skull, and scalp before they are picked up by the surface electrodes.

Scalp potentials are determined by the vectors of cortical activity. If the superficial layers of the cortex have a positive field potential and the deeper layers have a negative field potential, then the vector is vertical, with the positive end pointing toward the scalp electrodes. The amplitude of the vector depends on the total area of activated cortex and the degree of synchrony between cortical neurons. Scalp electrodes are not able to record electrical activity of deep nuclei. Their effective recording depth is only a few millimeters.

Basic Electroencephalographic Rhythms

EEG rhythms are classified into four frequency bands (Table 3.1). No individual band is normal or abnormal by definition. All are interpreted in the context of the topographic location and age and conscious state of the patient.

Alpha Rhythm

The alpha rhythm is usually seen in normal, relaxed individuals who are awake with their eyes closed. It is approximately 10 Hz in adults with the maximum voltage originating from the occipital electrodes, O1 and O2. The term *alpha rhythm* is used by some physiologists to signify any posterior dominant rhythm regardless of frequency, but this is an improper use of the term.

TABLE 3.1   Electroencephalography rhythms

Rhythm	Description	Normal	Abnormal
Alpha	8–13 Hz	Posterior dominant rhythm in older children and adults	Diffuse alpha in alpha coma Can signify seizure activity, especially in neonates
Beta	>13 Hz	Normal in sleep, especially in infants and young children	Drug-induced frontal beta Breach rhythm over a skull defect
Theta	4–7 Hz	Drowsiness and sleep Posterior slow waves of youth may have a theta component	Temporal theta in the elderly Focal theta over a structural lesion
Delta	<4 Hz	Sleep	Intermittent rhythmic delta activity Polymorphic delta activity with focal lesions
Spikes and sharp waves	Spike: 25–70 ms duration Sharp wave: 70–200 ms duration	Vertex waves and frontal sharp transients in neonates Positive occipital sharp transients of sleep Benign epileptiform transients of sleep 6/sec phantom spike and wave 14-and-6 positive spikes	Focal and generalized epileptiform activity

In children, the dominant posterior rhythm is slower and may not attain the minimal frequency of 8.5 Hz until 12 years of age. Slower frequencies in a 12-year-old would be interpreted as abnormal and would most likely indicate a diffuse encephalopathy or, if unilateral, suggest a structural lesion.

The posterior dominant rhythm is suppressed by eye opening and promptly returns when the eyes are closed. This reactivity of the posterior alpha rhythm should be routinely tested during EEG recording. The posterior rhythm is suppressed if the patient is tense during the recording. The lack of posterior rhythm should not be interpreted as abnormal in this situation. Other EEG features that suggest a tense state include frequent eye blinks and muscle artifact in frontal and temporal leads.

The amplitude of the posterior rhythm is 15–50 µV in young adults. Older individuals often have lower amplitude, but the frequency is the same. Low amplitude should not be interpreted as abnormal if the frequency composition is normal. Slowing of the posterior dominant rhythm is not a normal part of aging. Amplitude asymmetries of the dominant rhythm are common. The amplitude is usually higher from the nondominant hemisphere, but the difference should not exceed 50%.

A prominent alpha rhythm can be recorded during anesthesia and coma, but the distribution is different from the normal posterior alpha rhythm. In anesthesia and coma, the alpha rhythm is generalized with an anterior predominance. This alpha activity is invariant and monotonous, lacking the usual modulation in frequency and amplitude of an occipital alpha. The appearance of alpha coma in a patient signifies a poor prognosis for good neurologic recovery.

Beta Rhythms
EEG activity with frequencies faster than 13 Hz occurs in all individuals but is usually of low amplitude and often overlooked in favor of slower frequencies during wakefulness and sleep. Beta activity is normally distributed maximally over the frontal and central regions. A low-amplitude high-frequency beta is especially prominent during normal sleep in infants and children and is enhanced by several sedatives, especially barbiturates and benzodiazepines. In some children, the beta activity is so prominent as to dominate the record. People with hyperthyroidism may accelerate their posterior rhythm from 10 to 14 Hz or more. This is technically in the beta range, but the rhythm continues to react like an occipital, awake, resting rhythm, and should be considered no different than the alpha rhythm in this context.

Alterations in the frequency, amplitude, and abundance of beta activity should be commented on in the description of the record but interpreted with caution. Marked asymmetry in beta activity suggests the possibility of a structural lesion on the side lacking the beta. Focal, high-amplitude beta activity can be recorded over a skull defect, which might be a burr hole or fracture site. This is termed *breach rhythm*.

Theta Rhythms
EEG activity with a frequency between 4 Hz and 8 Hz is seen in normal drowsiness and sleep in adults. Young children may show theta activity during the waking state, which makes interpretation of encephalopathy particularly

difficult. Digital frequency analysis does reveal a small amount of theta in the waking EEG of adults, but the content is quite small and the amplitude is low. The detection of this theta usually requires high-sensitivity recordings or digital frequency analysis.

Posterior slow waves of youth may be in the theta or delta range. Theta activity in the temporal region in older individuals has been ascribed to vascular disease. While the significance of temporal theta is controversial, we suspect that it is not part of normal aging. We suggest commenting on the presence of temporal theta in the body of the report and interpreting it as a mild abnormality.

Delta Rhythms

Delta activity is not normally recorded in the awake adult but is a prominent feature of sleep and becomes increasingly abundant during the progress from stage 2 to stage 4 sleep. Focal polymorphic delta activity may be recorded over localized regions of cerebral damage. Intermittent rhythmic delta activity is recorded when there is dysfunction of the relays between the deep gray matter and cortex. The activity has a frontal predominance in adults and is called *frontal intermittent rhythmic delta activity*, while in children the activity has an occipital predominance and is called *occipital intermittent rhythmic delta activity* or *posterior intermittent rhythmic delta activity*.

Generation of Epileptiform Activity

Epileptiform activity is generated when depolarization of the cortex results in synchronous activation of many neurons. It is conceptually attractive to equate action potentials with EEG spikes, but action potentials occur normally. The abnormalities in epileptiform activity are the repetitive nature of the discharge and the degree of synchrony. By virtue of the nature of scalp EEG recordings, synchronous activation of many neurons is required for generation of both normal and abnormal rhythms.

Spikes and Sharp Waves

Epileptiform activity consists mainly of spikes and sharp waves. While epileptiform slow activity and suppression does occur, this is much less common. Spikes have a duration of less than 70 ms, while sharp waves have a duration of 70–200 ms.

A sustained depolarization of a neuron is associated with multiple action potentials on the crest of the depolarization. If one neuron is affected by this activation, the scalp electrodes cannot detect it. But when the depolarization spreads synchronously to many neurons through associative connections, the summed field potentials are detected on the surface as a spike, with negativity at the focus. The foundation for this bursting of spike discharges is the *paroxysmal depolarization shift* (discussed below).

Usually, the negative end of the epileptiform dipole points to the cortical surface. Therefore, most recorded spikes are scalp-negative. The distribution of the negativity across the scalp surface is the *field*. The field can be represented as a topographic map, as shown in Figure 3.2.

Spikes and sharp waves are occasionally surface positive. Positive sharp waves are seen in intraventricular hemorrhage in the newborn and in two

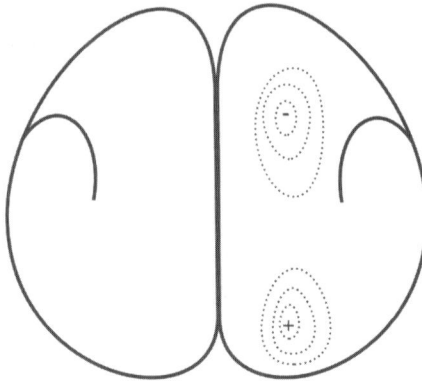

FIGURE 3.2 Sample topographic distribution of a spike focus. The negative end is in the right frontal region, whereas the positive pole is in the occipital region. The dipole is often oriented perpendicularly to the cortical surface so that the positive end of the dipole is not seen on the cortex.

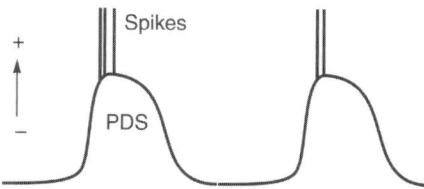

FIGURE 3.3 Paroxysmal depolarization shift. The paroxysmal depolarization shift is a wave of depolarization that can reach threshold, so that there are multiple spike discharges on the crest of the depolarization.

normal patterns: 14- and 6-Hz positive spikes and positive occipital sharp transients of sleep (POSTs).

Paroxysmal Depolarization Shift

Paroxysmal depolarization shifts (PDS) are extracellular field potentials that are waves of depolarization followed by repolarization (Figure 3.3). High-amplitude afferent input to the cortex produces depolarization of cortical neurons sufficient to trigger repetitive action potentials, which in turn contribute to the potential recorded at the surface. Repolarization due to inactivation of interneurons is followed by a brief period of hyperpolarization.

Cyclic depolarization and repolarization is believed to be the intracortical counterpart of the rhythmic spike activity seen in epilepsy. Rhythmicity is probably caused at least in part by the inability of cortical neurons to sustain prolonged high-frequency discharges. The loss of repetitive discharge is not caused by cellular exhaustion but by a built-in mechanism of inactivation after sustained discharge. Termination of the sustained depolarization is likely accomplished by activation of potassium channels and inactivation of the calcium channels, which are at least partly responsible for the prolonged depolarization.

Termination of the epileptiform activity is probably caused by inhibitory feedback to the neurons. The inhibitory neurons can exhibit a bursting similar to that seen in excitatory cortical neurons. Inhibition can suppress and ultimately stop the bursting that is feeding other neurons. Ultimately, the repetitive activation stops, often with a time of relative suppression.

Technical Aspects of Electroencephalography

Electroencephalographic Equipment

EEG has changed greatly in recent years. For many years, machines were fairly standard, analog machines with paper display. Sixteen channels was the norm, though additional channels helped with localization. Most of the recent advance has been in the conversion to digital machines. These usually use a cathode-ray tube (CRT) display, although there are a few older machines that would print on paper. Now, paper print is mainly used for only selected portions of the record rather than the entire record.

Digital vs. Analog Display

EEG equipment is either analog or digital. Analog machines are commonly in use, although digital machines are of increased popularity (Figure 3.4). Analog machines show a direct representation of the cerebral signal amplified many times to be displayed on paper. Analog filters change the frequency components of the displayed signal prior to writing on paper. Digital EEG machines produce display output that looks similar to the output of an analog machine, but the signal handling is very different. The cerebral signal is amplified, then digitized, as discussed in Chapter 2. Digital filters alter the frequency components of the signal. The display is usually on a computer monitor,

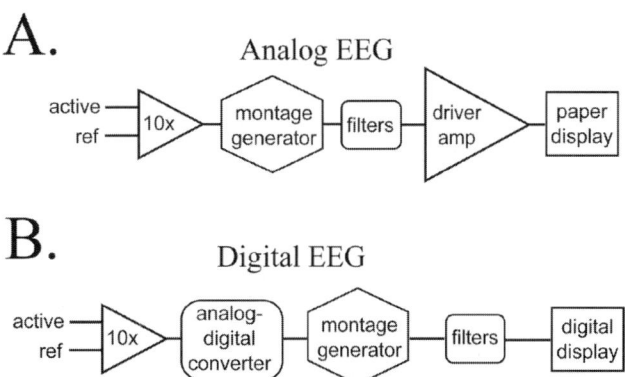

FIGURE 3.4 *Diagram of analog vs. digital EEG recording.* **A.** *Flow chart of the function of analog EEG. The initial 10× amplifier elevates the magnitude of the signal so that there is sufficient signal voltage to reduce apparent artifact and signal degradation.* **B.** *Flow chart of the function of digital EEG. The analog-to-digital converter, montage generator, and digital filters are part of the computer that is the core of the machine. The display may also be integral to the computer, but can also be an external, secondary high-resolution display.*

although the results can be written on paper. In the interest of conservation of paper and ease of interpretation, electronic display is preferable to paper display. Electronic display allows for alteration of filters, gain, and montage "on-the-fly."

Digital machines have several advantages over analog machines:

- Flexible display montage
- Continuously changeable gain
- Reduced storage assets
- Lack of mechanical distortion

Digital interpretation should be used only with caution and not as a substitute for visual interpretation. Quantitative differences may be significant mathematically yet meaningless clinically. Similarly, mathematical analysis may miss findings of great clinical importance. No hardware and software combination is close to threatening human interpretive abilities.

Display Montage Most digital machines record the potential from each electrode so that the montage can be altered by the reader. A spike seen on a longitudinal bipolar montage can then be viewed on a transverse bipolar montage. This can help localization, but the reader has to realize that multiple views of the same event do not represent multiple occurrences of the event.

Gain High-amplitude transients can peg the pens of an analog display, so some information is lost. Digital machines allow the gain to be changed, so that high and low-amplitude epochs will not be lost.

Storage Assets A busy EEG lab can generate tons of EEG paper. Although optical disks, CDs, and DVDs cost more to record, the amount of information that can be placed on a disk results in a lower cost for archiving the data. Unfortunately, not all of the EEG machines use the same format for storage of the digital data, so that reading the information in the future can require having compatible machines. This problem could be solved by having the data stored in a standardized format.

Mechanical Distortion EEG machines have mechanical distortion including arc distortion, inertial distortion, and overshoot. *Arc distortion* is due to the fact that the pen of a paper display moves around a pivot point, therefore, a perfectly vertical signal would produce an arc response (Figure 3.5). *Inertial distortion* is due to the fact that the pens have mass and pressure on the paper. Therefore, it takes more energy to get the pen moving vertically than it does to keep it moving. Although the paper is moving horizontally, getting it to move vertically requires more energy. The other extreme of inertial distortion is potential *overshoot* of the pen when it is moving vertically. In the absence of some corrective maneuvers, the pen overshoots its vertical target. Mechanisms to compensate for these mechanical distortions on paper recordings include electrical compensation of the amplifiers and adjusting pen pressure. *Damping* is the term used to describe the compensatory mechanisms

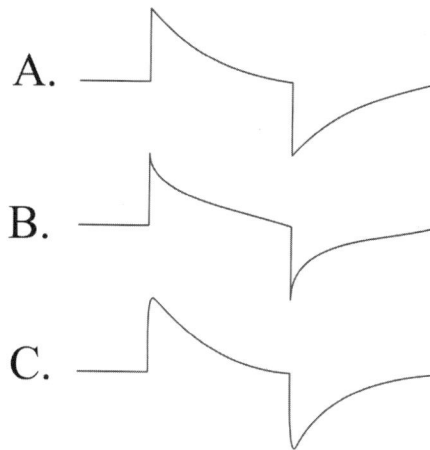

FIGURE 3.5 *Mechanical distortion.* **A.** *An ideal response of EEG amplifier, filter, and display to a square-wave calibration pulse.* **B.** *Mechanical distortion on paper because of overshoot of the pens. The inertia of the moving pen carries the recording beyond the target of the display. This error can be reduced by adjustment of pen pressure and by electrical corrective feedback.* **C.** *Mechanical distortion when the compensatory mechanisms described in B are excessive. The peak of the potential is rounded.*

for inertial distortion. Digital machines are immune to these mechanical distortions.

Number of Channels
The effective minimum number of channels for routine recording is eight; however, most neurophysiologists require 16. The use of 21 provides additional channels to montages, which can greatly help localization. Intensive care unit (ICU) monitoring for patients in status epilepticus can be performed with four channels; however, a baseline recording with a complete set of electrodes should be performed for evaluation of the discharge and background.

Factors to Consider in Purchasing Electroencephalographic Equipment
There are many factors to be considered before making a decision; however, a few factors deserve special mention.

Cost Cost of the machine is a small part of the expense of an EEG lab. The space and technical help are costs that dwarf the cost of the EEG machine. Therefore, it makes little sense to skimp on machine features for the sake of a small amount of money.

Technical Familiarity Technicians should ideally use equipment that functions similarly between facilities. Therefore, decisions about office machinery should consider equipment that is available or to be acquired by other offices and medical facilities where the technicians will work. Physicians also will be using equipment in different facilities, so identical or similar function is highly desirable.

Service EEG equipment occasionally needs service, so service needs to be considered. Guarantees of function and short-time to repair are important. Discussion of the local service facilities with other offices can be illuminating. An EEG lab that is nonfunctional for a week can lose a substantial part of the profit margin and referral base.

Portability At least one of the hospital EEG machines must be portable so that ICU, emergency room (ER), and operating room (OR) recordings can be made. There are special requirements for equipment that will be brought into the OR, but most modern machines meet these requirements. This has to be specified, however, if use in the OR is at all anticipated.

Reading Stations This is a moot point for paper EEG, but digital EEG can only be interpreted at workstations. Number, location, and availability of workstations should be considered. For example, in one of our hospitals, EEG data are retained on a drive in the machine, so when a portable study is being performed in the OR or ICU while disconnected from the network, EEGs performed on that machine cannot be read. This is not optimal.

Electrodes

Surface Electrodes

Surface electrodes are used for almost all routine EEG applications. They are disks that are fixed to the skin in a variety of ways. Electrode gel forms a malleable connection between the rigid disk and the skin. The electrode is secured on the scalp usually by pressing a gauze pad onto the gel–electrode combo. This can hold for a short time, providing that the patient is cooperative so that the head does not move. This electrode system does not work well if the patient moves or if the patient has stiff hair or other impediment to fixation of the electrode. Collodion fixation is a much more secure method of securing the electrodes on the scalp, although it does require special equipment, adequate ventilation, and more technician time. We usually use collodion preferentially in our laboratory even for routine outpatient studies.

Application of surface electrodes using gel consists of the following steps:

1. Locate the positions for electrodes using the 10–20 Electrode Placement System (explained later).
2. Separate strands of hair over the electrode positions using the wooden end of a cotton-tipped applicator.
3. Clean dead skin and dirt from the region with an agent such as Omni-Prep using the cotton-tipped applicator.
4. Scoop some gel into the electrode.
5. Place the electrode in position over the skin.
6. Put a 2″ × 2″ gauze pad over the electrode and push it firmly onto the head, providing a seal that prevents the electrode from falling off the scalp.

Application of electrodes with collodion involves the following steps:

1. Prepare the head at the electrode positions as mentioned for electrode gel.
2. Place the electrode on the scalp.
3. Place a piece of gauze soaked with collodion over the electrode.
4. Use compressed air to dry the collodion.
5. Insert a blunt-tipped needle into the cup and scrape the skin to lower electrode impedance.
6. Inject electrolyte into the cup of the electrode using the blunt-tipped needle.

Removal of electrodes is easy for the gel fixation. The gauze pads are pulled off, then the electrodes gently pulled off, tilting them to release the vacuum that holds them on. Then, the gel left on the scalp can be largely removed by rubbing with a warm, wet wash cloth. The patient can easily remove all traces of the gel by normal washing of the hair at home.

Collodion is more difficult to remove. First, the collodion is softened by use of acetone, then the areas cleaned as above. The degree of washing required is greater both immediately by the technician and later by the patient. Some patients object to the acetone smell more than any other part of the procedure.

Each method has its advantages. Collodion provides a more secure attachment and is more suitable for long-term recordings. Electrode gel is easier to apply and remove and is suitable for most routine office and hospital recordings.

Needle Electrodes

Needle electrodes offer no advantages over conventional surface electrodes and should not be used for routine studies unless recording cannot be accomplished any other way. The risk of infection to the patient and technician is unacceptably high.

Sphenoidal Electrodes

Sphenoidal electrodes are used to record activity from the temporal lobe that would not show on scalp recordings. The electrodes are inserted percutaneously adjacent to the zygoma until they reach the base of the skull. Sphenoidal electrodes should only be used by physicians trained in their insertion and experienced in interpretation of the recorded potentials.

Subdural Strip Electrodes

Subdural strip electrodes are used to evaluate patients for epilepsy surgery. The strips are placed during surgery through burr holes. The strips allow for a detailed map of the recorded electrical activity. Subdural strip electrodes should only be used by physicians trained and experienced in placement and interpretation, and only as part of a comprehensive epilepsy intervention program.

Depth Electrodes

Depth electrodes are used to localize seizure foci for surgery. A depth electrode consists of an array of electrodes on a single barrel that is inserted into the brain, usually in the temporal lobe. Only trained and experienced epileptologists should use depth electrodes.

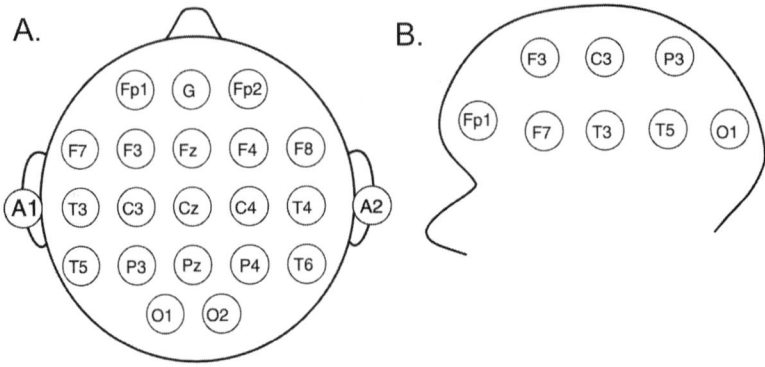

FIGURE 3.6 The 10–20 Electrode Placement System. **A.** Superior view. **B.** Left side view.

Electrode Position

Electrodes should be placed according to the 10–20 Electrode Placement System, as recommended by the International Federation of Societies for EEG and Clinical Neurophysiology. This system uses 21 electrodes placed at positions that are measured at 10% and 20% of head circumference (Figure 3.6).

Electrodes are named according to their position within the 10–20 system, with the first part of the name referring to the region and the second part referring to the area within the region. For regions, the following assignments have been made:

- F = frontal
- C = central
- P = parietal
- T = temporal
- O = occipital
- A = auricular (ear)
- Fp = frontopolar

The second part of the electrode name indicates the exact location. Numbers refer to pre-arranged locations, and the reader has to know the 10–20 system to know the difference between T3 and T5, for example. Odd numbers are on the left side of the brain, even numbers are on the right. Within a region, lower numbers are more anterior and medial to higher numbers. Midline electrodes are designated using "z" rather than a number.

Some examples of this numbering:

- Fz = frontal at the vertex
- Cz = central at the vertex
- T3 = left temporal
- T5 = left temporal, posterior to T3.

The head is measured in the following manner.

1. Measure the distance from the nasion to inion across the vertex. Mark a line 50% of this distance at the top of the head.
2. Measure the distance between the preauricular points, just in front of the ear. Mark a line at 50% of this distance, at the top of the head. The intersection of this line with that of step 1 is Cz.
3. Lay the measuring tape from nasion to inion through Cz. Mark 10% of this distance above the nasion for Fpz and above the inion for Oz. Fz is 20% of this distance above Fpz. Pz is 20% of this distance above Oz.
4. Lay the tape between the preauricular points through Cz. T3 is 10% of this distance above the left preauricular point and T4 is 10% of this distance above the right preauricular point. C3 is 20% of this distance above T3, and C4 is 20% of this distance above T4.
5. Lay the tape from Fpz to Oz through T3. Fp1 is 10% of this distance from Fpz, F7 is 20% of this distance posterior to Fp1. O1 is 10% of this distance anterior to Oz, and T5 is 20% of this distance anterior to O1. Measure in the same manner for Fp2, F8, O2, and T6 over the right hemisphere.
6. Lay the tape from Fp1 to O1 through C3. F3 is half the distance between Fp1 and C3. P3 is half the distance between C3 and O1. Repeat for the right side, with the tape from Fp2 to O2 through C4. F4 is half the distance between Fp2 and C4, and P4 is half the distance between C4 and O2.
7. Lay the tape from F7 to F8 through Fz, F3, and F4 to ensure that the distance between the electrodes is equal. Then lay the tape from T5 to T6 through Pz, P3, and P4 to ensure equal interelectrode distances.

Abbreviations for special electrodes and less standardized, and may differ between laboratories. *Sp* usually indicates sphenoidal electrode and *Naso* usually indicates nasopharyngeal electrode. Odd numbers are left-sided and even numbers are right-sided. Subdural strip and depth electrodes are also named using letters and numbers where the letter indicates the array and the number indicates which electrode in that array.

Montages

The sequence of electrodes being recorded at one time is called a *montage*. All montages fall into one of two categories: bipolar or referential. *Referential* means that the reference for each electrode is in common with the other electrodes. The reference could be a single noncephalic electrode, ipsilateral

TABLE 3.2 Left parasagittal portion of the longitudinal bipolar montage

Channel number	Active electrode	Reference electrode	Output value
1	Fp1	F3	Fp1-F3
2	F3	C3	F3-C3
3	C3	P3	C3-P3
4	P3	O1	P3-O1

ear, or perhaps even Cz, though this is not a useful reference because of its electrical activity. *Bipolar* montage means that the reference electrode for one channel is the active electrode for the next channel. Bipolar montages are particularly useful for visual analysis of focal cerebral activity such as spikes and sharp waves. For example, in the *longitudinal bipolar (LB)* montage, the first four channels are shown in Table 3.2.

For all channels, negativity at the active electrode produces an upward deflection of the pen on the paper. Negativity at the reference produces a downward deflection. The *Guidelines* recommends the following principles in designing montages:

- Record at least eight channels.
- Use the full 21 electrode placement of the 10–20 system.
- Every routine recording session should include at least one montage from each of the following groups: referential, longitudinal bipolar, and transverse bipolar.
- Label each montage on the recording.
- Use simple montages that allow for easy visualization of the spatial orientation of waveforms—for example, bipolar montages should be in straight lines with equal interelectrode distances.
- Have the anterior and left-sided channels above the posterior and right-sided channels.
- Use at least some montages that are commonly used in other laboratories.
- Remember that negativity in the active electrode of each channel produces an upward deflection of the pen.

The recommended montages for routine use in adults are shown in Table 3.3, and in Figure 3.7. Additional channels, when available, are used

TABLE 3.3 Recommended electroencephalographic montages

Channel	LB	TB	Ave	Ref
1	Fp1-F3	Fp1-Fp2	Fp1-Ave	Fp1-A1
2	F3-C3	F7-F3	Fp2-Ave	Fp2-A2
3	C3-P3	F3-Fz	F3-Ave	F3-A1
4	P3-O1	Fz-F4	F4-Ave	F4-A2
5	Fp2-F4	F4-F8	C3-Ave	C3-A1
6	F4-C4	A1-T3	C4-Ave	C4-A2
7	C4-P4	T3-C3	P3-Ave	P3-A1
8	P4-O2	C3-Cz	P4-Ave	P4-A2
9	Fp1-F7	Cz-C4	O1-Ave	O1-A1
10	F7-T3	C4-T4	O2-Ave	O2-A2
11	T3-T5	T4-A2	F7-Ave	F7-A1
12	T5-O1	T5-P3	F8-Ave	F8-A2
13	Fp2-F8	P3-Pz	T3-Ave	T3-A1
14	F8-T4	Pz-P4	T4-Ave	T4-A2
15	T4-T6	P4-T6	T5-Ave	T5-A1
16	T6-O2	O1-O2	T6-Ave	T6-A2
17	Fz-Cz	Fz-Cz	Fz-Ave	Fz-A1
18	Cz-Pz	Cz-Pz	Cz-Ave	Cz-A2

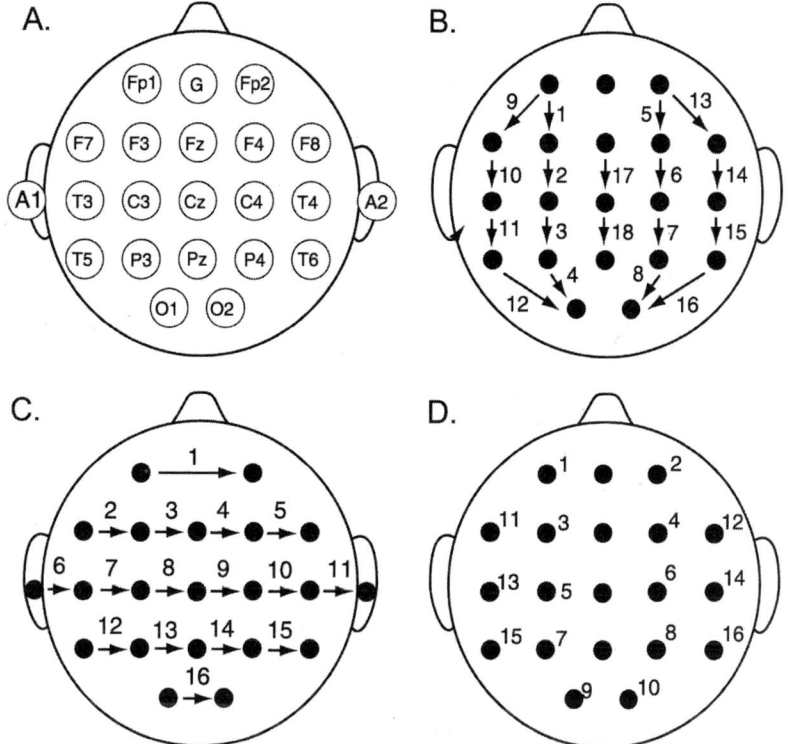

FIGURE 3.7 *Bipolar and referential montages. Common montages used in routine EEG.* **A.** *Electrode positions.* **B.** *Longitudinal bipolar (LB) montage.* **C.** *Transverse bipolar (TB) montage.* **D.** *Referential montage.*

for monitoring other biological functions, such as ECG, eye movements, respirations, and EMG.

Routine Electroencephalography

EEG recordings are usually stored on paper or disk. There is a face sheet that is attached to the paper record, if one exists, or is separate, in the case of digital recordings. For all recordings, the face sheet has all of the basic identifying information:

- Name
- Age
- Identification number of the patient
- Index number of the recording
- Reason for the study
- Name of the technician
- Current medications
- Time of the last seizure, if appropriate
- Technical summary, including activation methods and artifacts

- Technician's observations, including regions of particular concern or interest
- Time and date of the recording
- Ordering clinician
- Sedative medication used

Digital recordings should be identified on the face sheet, with the index number of the disk and the format of storage.

The face sheet should be filled out completely before the patient leaves the lab. If the technician sees a finding of immediate clinical concern, the neurophysiologist should be called immediately.

Calibration

Two phases of calibration are used prior to each study—square-wave calibration and biological calibration.

Square-Wave Calibration A square-wave pulse is delivered from a waveform generator into each amplifier input. This pulse is 50 µV in amplitude and alternated on and off at 1 second intervals. The wave does not appear precisely square because of the effects of the preset default filters. Figure 3.8 shows a sample recording.

The low-frequency filter (LFF) transforms the plateau of the square wave into an exponential decay. The high-frequency filter (HFF) slightly rounds off the peak of the calibration. For educational purposes, try several HFF and LFF settings during the calibration test to see the effects of filter changes on the record. It is also instructive to change filter settings during recording of EEG activity at a time that would not interfere with clinical interpretation.

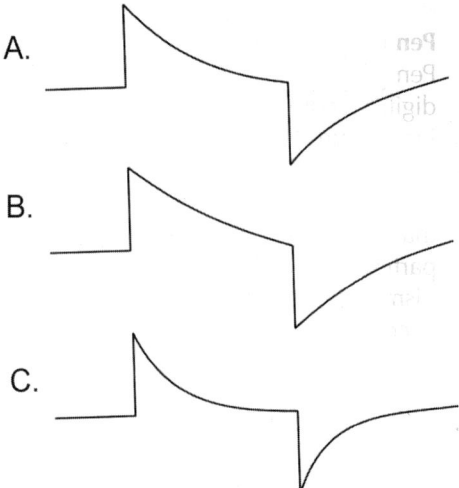

FIGURE 3.8 *Square-wave calibration signal responses with different filter settings.* **A.** *Standard filter settings.* **B.** *Reduction in the frequency setting of the low-frequency filters.* **C.** *Increase in the frequency settings of the high-frequency filter.*

The time constant (TC) of the LFF can be measured from the square-wave calibration page. TC is equal to the time it takes for a potential to fall to 37%, or roughly one-third, of peak value.

It is difficult to estimate the HFF setting from the square-wave calibration; however, neurophysiologists should have an idea of what the peak should look like. If the HFF is set too low, there will be a slow roll-off on the peak of the calibration pulse. If the HFF is set too high, the wave will appear too peaked and may even show overshoot, as if there is too little pen damping.

The experienced neurophysiologist can tell if there is a problem with one or more amplifiers by looking at the recordings made from each channel. Abnormalities are obvious if present in one channel; abnormalities in multiple channels may be more subtle. Types of abnormalities can include:

- Peak too rounded
- Peak overshoot
- Incorrect rate of decay

If the peak is too rounded, the high-frequency response of the system is impaired. This usually means that there is excessive pen damping or the high-frequency filter setting is improperly set. If the peak overshoots, pen damping is usually set too low, allowing for inertia to produce excessive pen excursion. Incorrect rate of decay can mean that a filter setting is wrong for one or more channels or that the amplifier is failing. Amplifiers may fail to the point that they give no response, but more commonly, they give distorted responses prior to failing completely.

Biological Calibration Biological calibration, or *Biocal*, assesses the response of the amplifiers, filters, and the recording apparatus on a complex biological signal. Electrodes Fp1 and O2 are connected to all of the amplifier inputs. The recordings from all of the channels should be identical.

Pen Pressure and Damping

Pen pressure and damping are issues that apply only to paper recordings; digital recordings are immune to these causes of distortion. Mechanical writing instruments have two inherent limitations: inertia and friction. Even when the filters are set properly, the frequency response may be inaccurate because of these mechanical factors. The physical mass of the pen produces inertia that slows its response time to sudden changes of signal voltage. Inertia is partially compensated for by control mechanisms in the pen drive mechanism. Friction is also compensated for by EEG machine electronics, but excessive pressure of the pen on the paper results in a sluggish response. This is the main reason that neurophysiologists and technicians need a visual memory of what a calibration pulse looks like with proper filter settings.

The same inertia that inhibits pen movement also promotes excessive pen movement, *overshoot*. Overshoot is minimized by the pen control mechanism, termed *damping*. When damping is not sufficient, normal waveforms may look like spike discharges. These effects are minimized with proper setting of pen pressure and damping. The manuals provided with EEG machines give instructions on setting of damping and pen pressure.

Sensitivity

The recording sensitivity is initially set at 7 µV/mm and subsequently adjusted depending on the amplitude of the EEG activity. Movement artifact and other noncerebral transients may exceed maximal pen excursion, but electrocerebral activity may not. Important waveforms may be missed when the sensitivity is set too low.

For children, the sensitivity is often reduced to 10–15 µV/mm because EEG amplitude is high in both the awake and sleep states. The elderly often have low-voltage EEG activity, and increased sensitivity is required. Studies performed for the determination of brain death are started at 7 µV/mm but the sensitivity is always increased to 2 µV/mm.

Duration

A routine EEG should include at least 20 minutes of relatively artifact-free record. Longer duration recordings are often helpful in neonates so that the transitions between states can be identified. The *Guidelines* recommend 30 minutes of recording for brain death studies.

Filters

The standard filter settings for routine EEG are:

- LFF = 1 Hz
- HFF = 70 Hz

The LFF of 1 Hz corresponds to a TC of 0.16 sec. If the LFF is set higher than 1 Hz, there will be attenuation and distortion of some slow waves. Slow waves have an increased number of phases and are composed of faster frequencies. Technicians should be discouraged from turning up the LFF especially when there is an abundance of slow activity. If the HFF is set too low, fast activity is blunted, and spikes and sharp waves may be impossible to identify.

The 60-Hz filter should not be needed in most laboratories. Three ways to minimize 60-Hz activity are:

- Selection of equipment location
- Grounding
- Shielding

Shielding of the room is desirable but usually not essential and cannot completely abolish artifact induced by strong electromagnetic fields. Studies in a special care unit usually require use of the 60-Hz filter. Sources of artifact include ventilators, intravenous infusion pumps, air beds, heating or cooling blankets, and monitoring equipment.

Activation Methods

The performance and interpretation of records obtained with activation methods are discussed in Chapter 4. Hyperventilation, photic stimulation, and sleep may activate epileptiform activity. After an initial period of recording in the relaxed, wakeful state, the patient is asked to hyperventilate for 3 minutes. If absence seizures are suspected, the patient is asked to hyperventilate for 5 minutes. Hyperventilation is not performed in elderly individuals or in patients with advanced atherosclerotic disease because of concern for vasoconstriction with resultant cardiac or cerebral hypoperfusion.

Photic stimulation is performed on older children and adults of all ages. Photic stimulation of sleeping infants is probably of limited clinical value.

Sleep is not considered by some to be a true activation method, because it is a transition between natural states. However, sleep helps to promote epileptiform activity, and in routine EEG, sleep frequently has to be induced by sedatives. Sleep deprivation may be needed if sedated sleep is not obtained. In this sense, sleep is an activation method. Sleep recordings are routinely indicated for all patients being evaluated for seizures, but are not helpful for patients being evaluated for encephalopathy. The mechanism of sleep is not important. There is no convincing evidence that natural sleep, sedated sleep, and sleep deprivation differ in their ability to promote epileptiform activity.

Electrode Impedance

Electrode impedance should be at least 100 ohms and no more than 5 kohm. Lower impedances cannot be obtained with scalp electrodes without there being some improper connection between the electrodes, such as smeared gel between adjacent electrodes. Higher electrode impedances can create impedance mismatch, which can then degrade the rejection of artifact by bipolar recordings.

Electrical interference is common in EEG laboratories, but since most recordings are bipolar, even referential recordings, the noise is typically similar in conformation and amplitude in the leads. Since the bipolar recording setup subtracts the voltage of the reference from the active signal voltage, the noise cancels out. A signal that is common to the two inputs is called *common mode*. This rejection can be described by the *common mode rejection ratio*. This represents the ability of an amplifier to reject signals common to two inputs of the amplifier.

Telephone Transmission Electroencephalography

Telephone transmission of EEG is useful and accurate with certain caveats. The interpreter must be aware of several sources of artifact from public telephone lines and the equipment required to transform the data into digital form for transmission. The availability of high-speed internet access now allows for transfer of information, although there are special considerations for transfer of patient data over the internet.

The *Guidelines* recommend that telephone transmission EEGs be performed in accordance with the guidelines for routine EEGs. In addition, they make the following recommendations:

- Manufacturers of the equipment must provide specifications on frequency response, noise and crosstalk. The equipment must be checked periodically to ensure that the specifications are continually met.

- The equipment should indicate if there is difficulty with transmission or reception.
- Integrity of transmission should be checked before and after each recording.
- The record should be labeled as for any recording. In addition, there should be designation that this is a telephone transmission recording.
- A paper recording should be made at both the transmitting and receiving stations, for accurate relay of information on physiologic state, activity, and artifacts.
- The EEG from both transmitting and receiving stations should be stored for future comparison.
- The technicians at the transmitting and receiving stations should be well trained not only in routine EEG but also in the techniques and problems associated with telephone transmission EEGs.
- Telephone transmission EEG cannot be used as a confirmatory test for determination of brain death.

Electrode caps are commonly used in facilities that seldom perform EEGs. The technical staff usually do not perform EEGs as their primary duty, so the caps offer efficiency and relatively consistent electrode placement.

In general, most neurologists recommend use of free electrode placement following careful head measurement. However, rural areas without technicians who perform studies frequently are less likely to make errors of electrode placement with the use of the electrode caps. There is the potential for error, however. We have more than once found that the cap was placed backwards on the head at the outside laboratory.

Technicians at outside facilities are seldom as well trained and supervised as technicians in the home EEG laboratory. Therefore, there is greater opportunity for technical error. Many outside facilities use electrode caps that fit snugly on the head. There are several problems with electrode caps:

- Impedances are usually higher than with normally applied electrodes
- The cap may not be positioned perfectly on the head
- The electrode positions of the electrodes differ between patients because of head size.

Electroencephalography Laboratory

EEG can be performed in almost any part of the hospital, but routine studies are performed in the EEG lab, where electrical and acoustic noise is minimal, and all ancillary equipment is available. Many labs have the patient in a separate room from the recording equipment and technician, but this is not usually necessary for routine daytime recordings.

The EEG laboratory should conform to standard guidelines for electrical safety, including three-prong plugs with proper grounding. Shielding is usually not needed with proper grounding, good technique, and bipolar montages. However, if the EEG laboratory is forced to be in an electrically noisy location, electrical shielding may be necessary. If the laboratory is in an acoustically noisy location, acoustic insulation may be necessary. These factors greatly increase the cost of the laboratory.

Report

The EEG report should be clear and concise, and ideally be no longer than one page. Sample recordings are seldom attached to EEG reports, as they may be for EP and EMG reports. The three fundamental components to the report are:

- Report header
- Description of the record
- Interpretation

Report Header
The report header has all of the identifying information, including:

Patient information

- Name of the patient
- Date of birth
- Sex
- Identification number

Laboratory information

- Name of laboratory
- Address and phone number of laboratory

Study information

- Date of study
- Reason for study
- Referring physician
- Interpreting physician

Description of the Record
The description of the record should be complete without containing exhaustive detail. Among the information included in the description of the record is:

- State and state changes—awake, drowsy, sleep—and description of the findings at each level.
- Background rhythms—posterior dominant rhythm, frontal rhythms, temporal theta.

- Focal abnormalities—asymmetries, focal slowing, focal sharp activity.
- Epileptiform activity—spikes and sharp waves whether focal or generalized, and regional prominence.
- Artifacts that interfere with interpretation of the record, and impression whether the study is interpretable.

Interpretation

The interpretation is the most important part of the report. While there is considerable room for personal preferences, the following information must be provided:

- Normal or abnormal
- How the recording is abnormal
- Clinical implications of the findings.

For example, some sample interpretations might be:

"Normal awake and sleep EEG"

or:

"Abnormal study because of generalized three-per-second spike–wave complexes. This is consistent with a seizure disorder of the generalized type."

The clinical interpretation should take into account the information provided by the ordering physician. For example, if the clinical history is of partial complex seizures and the EEG shows a spike focus, the interpretation might be:

"Abnormal study because of a spike focus in the right anterior temporal region. This is consistent with a partial seizure disorder."

On the other hand, if the clinical history is of a behavioral disorder, the impression might be:

"Abnormal study consistent with a sharp-wave focus in the left posterior-temporal region. This could be consistent with a seizure disorder, but patients without seizures can manifest this pattern."

Record Keeping

Printed EEG reports should be kept for the duration of the practice, although there are no clear published recommendations on this issue. Patients with seizures may live entire lives with multiple studies performed over time, and a comparison can be extremely helpful.

The paper EEG record cannot be kept forever, but at least parts of the record should be kept for at least 2 years. Microfilming can greatly improve record keeping of paper records. Digital recording has allowed for virtually indefinite storage of the complete record, and will be the predominant method of storage in the future. Optical disk, CD, and DVD recordings have a duration and reliability that certainly exceed routine storage requirements.

State and local regulations may have additional requirements for record keeping.

4 Normal Electroencephalographic Patterns

Electroencephalography in Adults

Waking Record

Fast frequencies dominate the normal adult waking electroencephalogram (EEG). The posterior rhythm is recorded in the awake state with the eyes closed. Eye opening attenuates the posterior dominant rhythm. There is an anterior–posterior gradient, with faster frequencies seen over the frontal regions, usually in the beta range (Figure 4.1). These faster frequencies are of low amplitude, however.

Anterior Cerebral Activity
The background activity of the frontal region is composed predominantly of low-voltage fast activity with superimposed eye movement artifact. Rhythmic beta may be recorded from the frontal and central regions, especially when sedatives are used. Drug-enhanced beta is more commonly seen after benzodiazepine and barbiturate sedation than following chloral hydrate.

Theta and delta activity are not prominent in the normal awake adult EEG. However, digital EEG signal analysis shows a small amount of bihemispheric theta in most patients.

Posterior Cerebral Activity
In the awake state with the eyes closed, the predominant rhythm from the occipital region is about 10 Hz. The range of normal is 8.5–11 Hz, with frequencies slower than 8.5 Hz being abnormal, usually indicating dementia or encephalopathy. There is some oscillation of this rhythm, but it still has a very regular appearance.

The posterior dominant rhythm is usually symmetric, but asymmetries of up to 25% are frequently seen. Asymmetry should not be interpreted as abnormal unless it is at least 50%. The amplitude from the left hemisphere is often lower than the right, so this should be considered when interpreting amplitude asymmetries.

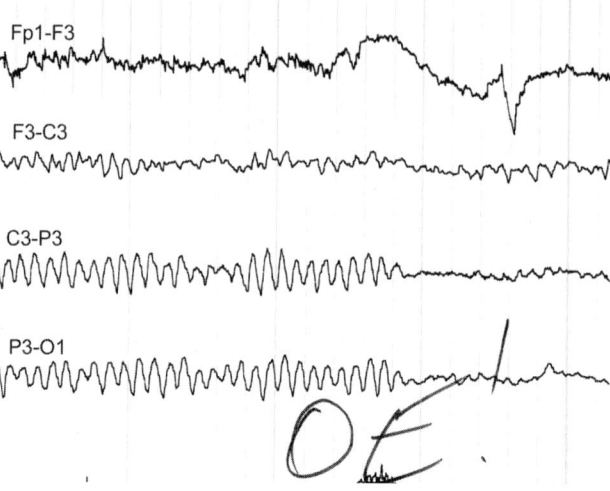

FIGURE 4.1 *Normal waking EEG recorded using the left parasaggital portion of the longitudinal bipolar montage. There is a posterior dominant rhythm of 9–10 Hz that attenuates to eye opening (signified as "OE!").*

Increasing age results in the posterior dominant rhythm being lower amplitude and less well organized. While slowing of the background is common in the elderly, it is still abnormal. We use a rigid cutoff of 8 Hz; any frequency less than this is abnormal at any adult age.

Tense state can suppress the posterior dominant rhythm even with the eyes closed, so relaxation is important to recording an adequate response.

Drowsiness (Sleep Stages 1A and 1B)

As the patient progresses from the relaxed awake state to drowsiness, the first sign is attenuation of the posterior dominant rhythm with some spread anteriorly, plus slight slowing of the background. This is stage 1A of sleep. As the patient sinks further into stage 1B, the posterior dominant rhythm is reduced to less than 20% with theta activity becoming more prominent.

Vertex waves may be seen at stage 1B, but are more a sign of stage 2 sleep. Differentiation of stages 1A and 1B is not important for routine EEG.

Sleep

Sleep Patterns

Background Rhythms   During drowsiness the posterior dominant rhythm is lost and replaced by a mixture of frequencies. Theta and delta appear with faster frequencies superimposed. The details of how the frequency compositions differ between the stages of drowsiness and sleep are discussed below.

Vertex Waves   Vertex waves are surface-negative potentials with a maximum amplitude on either side of the midline (C3 and C4). They are most common in stage 2 sleep and often appear at times of partial arousal. The vertex waves are a biphasic sharp wave with an initial negative deflection followed by a positive deflection. The sharp wave may be followed by a slow wave or a spindle, the latter being termed a *K complex*.

The vertex waves can be asymmetric, especially in children, and high amplitude in younger patients. The asymmetry should be abnormal if more than 25% and if consistent between the hemispheres. Vertex waves first appear at approximately 8 weeks of age. In children, vertex waves that appear in trains may be mistaken for seizure activity.

Sleep Spindles Sleep spindles are rhythmic 11–14 Hz waves whose duration is typically 1–2 seconds, with a minimum of 0.5 seconds, and whose amplitude is at least 25 µV. They are most prominent in the central regions during stage 2 sleep. Unlike vertex waves, the maximum amplitude of sleep spindles is typically seen lateral to the midline (C3 and C4). Asymmetry in the abundance of sleep spindles is normal unless sleep spindles fail to appear from one hemisphere.

Sleep Stages

Sleep stages are discussed below, and summarized in Tables 4.1 and 4.2.

Stage 1 Stage 1 is drowsiness, and is divided into stage 1A and 1B. Stage 1A is characterized by attenuation of the posterior dominant rhythm and some change in field of distribution so that it is seen more anteriorally than in the fully awake state.

Stage 1B is characterized by progressive loss of the posterior dominant rhythm, with less than 20% of the background composed of the alpha rhythm. Theta activity becomes more prominent. Vertex waves may be present in stage 1B; however, this is more commonly seen in stage 2. Differentiation between stages 1A and 1B is not important for routine EEG.

Stage 2 Stage 2 sleep is characterized by sleep spindles, vertex waves, increased theta, and the appearance of delta. However, less than 20% of the record contains delta. Since some vertex waves can be seen in stage 1B, the primary differentiating feature is the appearance of sleep spindles.

Stage 3 Stage 3 sleep is characterized by increasing delta content and reduction in faster frequencies. Delta constitutes 20–50% of the record.

TABLE 4.1 Sleep rhythms

Pattern	Description
Vertex waves	Negative potentials with a maximum at Cz
	Occur in stage 2 sleep and during arousal
Sleep spindles	11–14 Hz waves of 1–2 sec duration
	Maximum at C3 and C4
	Most prominent in stage 2 sleep
K complexes	Fusion of a vertex wave and a sleep spindle
	Prominent in stage 2 sleep and partial arousal
Positive sharp transients of sleep	Positive potentials with a maximum at O1 and O2

TABLE 4.2 Sleep stages

Stage	Features
1. Drowsiness	Attenuation of alpha, slight slowing, spread of alpha anteriorly 1A = light drowsiness 1B = deep drowsiness
2. Light sleep	Sleep spindles, vertex waves, increased theta activity, and the appearance of some delta
3. Slow-wave sleep	More delta activity, constituting 20–50% of the record Sleep spindles and vertex waves are less prominent
4. Slow-wave sleep	More delta activity, constituting >50% of the record Sleep spindles and vertex waves are often absent

Stage 4 Stage 4 sleep is characterized by a further increase in delta content, so that delta constitutes more than 50% of the record. Vertex waves and sleep spindles are less prominent and are often absent.

REM Rapid eye movement (REM) sleep is characterized by a low-voltage background composed of predominantly fast frequencies. It can be difficult to distinguish REM sleep from light drowsiness. Rhythmic 6–8 Hz activity, which is called sawtooth waves because of its unusual morphology, may appear in the frontal regions and the vertex.

Typically, REM sleep follows after progression from sleep stages 1 through 4. Progression from drowsiness to REM sleep without passing through other stages, termed *REM-onset sleep*, occurs in the following conditions:

- Narcolepsy
- After sleep deprivation
- After alcohol- or drug-induced REM deprivation sleep.

The features of REM sleep that distinguish it from drowsiness are the following:

- Rapid and chaotic eye movements (drowsiness has slow roving eye movements)
- Hypotonia on submental electromyography (EMG)
- Irregular respiratory rate.

Sequence of Sleep Stages Patients progress from relaxed wakefulness through stage 1A and 1B into stage 2 sleep. During a routine office EEG, deeper stages of sleep are seldom seen. In fact, when seen they may be misinterpreted as abnormal rhythms. Progression into slow-wave sleep, stages 3 and 4, occurs later. The patient descends into deeper stages of sleep then ascends to drowsiness or wakefulness during a night's sleep. The cycle is repeated three to four times per night, although all stages may not be seen in all cycles.

REM sleep occurs usually after at least one sleep cycle and increases in duration with subsequent cycles. Respiratory pattern is usually regular in all sleep stages except REM. Submental EMG activity declines progressively as sleep becomes deeper, until disappearing during REM sleep.

Electroencephalography in Children

Waking Electroencephalography

Maturation of the Posterior Rhythm
The posterior dominant rhythm in the awake infant is approximately 4 Hz. The rhythm becomes faster with age, reaching the normal adult frequency of approximately 10 Hz by 10 years. This maturation is shown in Figure 4.2. The amplitude gradually increases, such that by age 10 years the alpha is often in the range of 50–100 µV. In adults, the alpha amplitude gradually declines with increasing age.

Posterior Slow Waves of Youth
Slow waves of youth are seen predominantly in the waking state and occasionally in light drowsiness, stage 1. They are in the delta range and are superimposed on the normal posterior dominant rhythm. Slow waves of youth may be differentiated from pathologic slow waves by the otherwise normal background and their reactivity to eye opening. Slow waves of youth decrease with increasing age and are not seen after the age of 30 years.

Sleep Electroencephalography

Sleep promotes some epileptiform activity and is used as an activation method. However, the morphology of the epileptiform discharges may look different in sleep than they look in the waking state. Sleep patterns differ in children and adults.

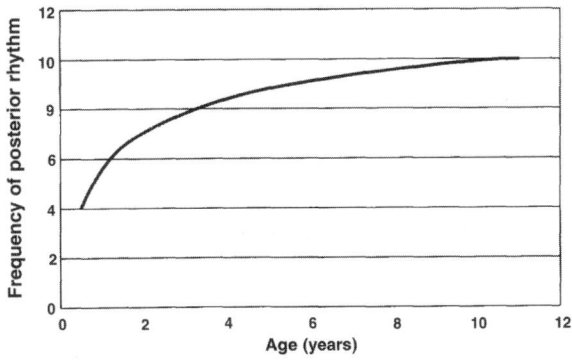

FIGURE 4.2 Maturation of the posterior dominant rhythm. The posterior dominant rhythm gradually increases during childhood years, reaching adult frequencies by 4–5 years of age. However, there are other differences that differentiate a child's EEG from an adult EEG.

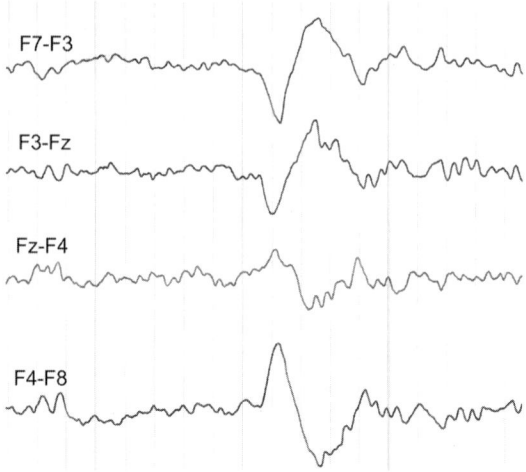

FIGURE 4.3 *Vertex wave during stage 2 sleep. This is the frontal portion of the transverse bipolar montage. The vertex wave has maximum negativity at Fz in this limited display.*

The posterior dominant rhythm of relaxed waking state attenuates with drowsiness. As the alpha disappears, theta activity appears from both hemispheres. This is commonly referred to as stage 1 sleep.

Vertex Waves

Vertex waves in children appear at about the age of 5 months. They are initially blunted in morphology (Figure 4.3). By 2 years of age the vertex waves are prominent, sharp in configuration, and high in amplitude. Clusters of vertex waves may be mistaken for epileptiform activity, especially since

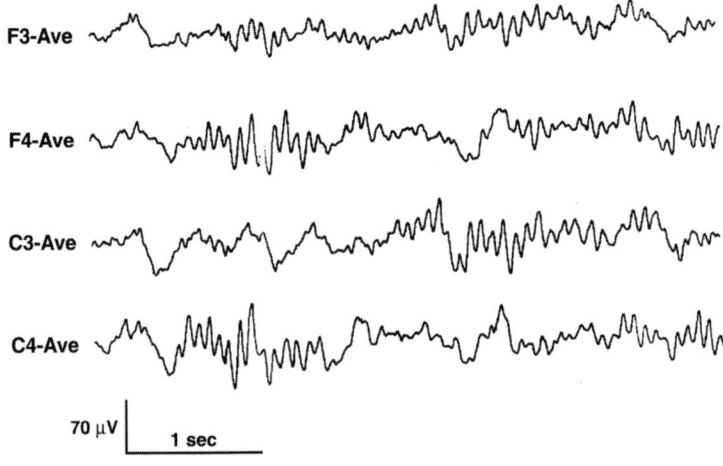

FIGURE 4.4 *Sleep spindles during stage 2 sleep. This is a portion of an average reference montage. The spindles in this example are independent on the two sides of the head.*

they can occur in trains and can be asymmetric, and can appear apart from the midline.

Sleep Spindles
Sleep spindles first make their appearance at about the age of 2 months. They are often prolonged and asymmetric. By 2 years, the sleep spindles have an adult appearance (Figure 4.4).

Normal Transients and Variants

Some common normal transients and variants are presented, and summarized in Table 4.3.

TABLE 4.3   Normal transients and variants

Waveform	Description	Features
Lambda	Waking record, when viewing a scene Positive occipital waves, blocked by eye closure	Completely normal
Mu	Waking rhythm Negative wicket-shaped spikes at about 10 Hz for about 1 sec Maximal at about C3 and C4	Blocked by movement of the contralateral extremity
Slow alpha variant	Waking record Notched posterior rhythm at 4–5 Hz Blocked by eye opening	Subharmonic of the posterior alpha Completely normal, though deserves comment
Wicket spikes	Drowsiness and light sleep Sharply contoured waves from the temporal region	More common with increasing age No following slow wave
14-and-6 positive spikes	Drowsiness and light sleep Sharply contoured positive rhythm at 14 or 6 Hz	Normal pattern, but increased by some disorders
Rhythmic mid-temporal theta of drowsiness	Drowsiness or relaxed wakefulness Temporal theta	Usually normal, but can be associated with structural lesions
SREDA	Awake pattern in older adults Rhythmic theta activity Looks different than seizure discharges	Most consider this to be normal, but is uncommon enough that comment should be made for later consideration
Mittens	Sleep pattern Appearance of a mitten with a slow wave bordered by a spike-like component	Formed by fusion of a sleep spindle with a vertex wave Clearly normal
Positive occipital sharp transients of sleep	Sleep pattern Irregular rhythm of positive sharp waves	Normal, no pathologic correlate

Lambda Waves

Lambda waves are normal positive waves that appear over the occipital region when the patient is looking at a picture or pattern. Lambda waves indicate visual exploration and are blocked by eye closure. They are called lambda waves because of their resemblance to the Greek lowercase letter lambda (λ).

Mu Rhythm

Mu rhythm is not a common feature of normal EEGs. It is a run of negative wicket-shaped spikes with an approximate frequency of 10 Hz and a duration from less than 1 second to many seconds. The negativity is maximal in the rolandic regions, mainly C3 and C4. Mu has the appearance of a centrally located alpha rhythm but is usually slightly faster than the patient's alpha rhythm. It is blocked by movement of the contralateral extremity; this is the key to identification. In fact, mu can be blocked by merely thinking about limb movement. The technician should ask the patient to move a limb to verify the waveform.

POSTS

Positive occipital sharp transients of sleep (POSTS) are surface positive potentials with maxima at O1 and O2 (Figure 4.5). They may occur as single waves or in trains. They resemble lambda waves except that they are present only in the sleeping state, whereas lambda waves are seen only in the waking state and with the eyes open.

POSTS may be related to replay of visual information during sleep, but this hypothesis is not universally accepted. They are not seen in patients who are blind or are severely visually impaired.

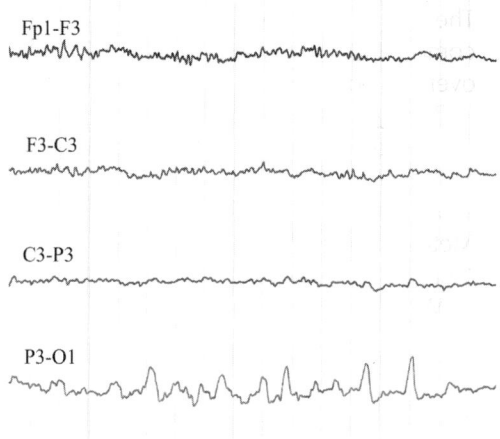

FIGURE 4.5   POSTS. Positive occipital sharp transients of sleep (POSTS) recorded using the left parasaggital portion of the longitudinal bipolar montage. The upward deflections in the last channel are due to positive potentials from the O1 electrode.

POSTS are not a consistent feature of sleep and have no diagnostic significance, unless they are seen only from one side. This asymmetry is unlikely to be the only sign of focal abnormality in an EEG record.

BETS

Benign epileptiform transients of sleep (BETS) are very small spike-like potentials that occur in the temporal regions during drowsiness and light sleep. They are less than 50 µV with a duration of less than 15 ms. Also called small sharp spikes, BETS are differentiated from epileptiform spikes by their small amplitude, short duration, lack of a slow wave, and normal EEG background.

Wicket Spikes

Wicket spikes are sharply contoured waves that are most prominent from the temporal regions during drowsiness and light sleep. They are differentiated from true spikes by:

- Absence of a following slow wave
- Normal background activity
- Occurrence in a series of waves at 6–10/sec

Wicket spikes are more common with increasing age.

14-and-6 Positive Spikes

The 14-and-6 positive spike rhythm is a sharply contoured positive waveform that occurs mainly in drowsiness and light sleep. The name comes from its variable frequency; sometimes it is a 14-Hz rhythm and sometimes it is at 6 Hz. A prolonged recording may be needed to see both frequencies, and both frequencies may not be seen in a single patient in a single recording. The 6-Hz pattern predominates in young children, whereas the 14-Hz component predominates in older children. This pattern is most prominent over the posterior temporal region, but has a widespread distribution.

The 14-and-6 positive spike rhythm has been associated with a number of pathologic conditions, but the association is generally weak and the incidence is likely not different from that in the general population. Therefore, this should be considered a normal variant, if the background is otherwise normal. Metabolic encephalopathies, such as hepatic encephalopathy, may have an increased incidence of this pattern, but the background is abnormal, as well.

When this pattern is encountered in interpretation of an EEG, the existence should be mentioned in the body of the report but should not be interpreted as abnormal, in the absence of any other findings.

Slow Alpha Variant

The *slow alpha variant* is a subharmonic of the normal posterior dominant alpha rhythm. The frequency is 4–5 Hz rather than the usual 10 Hz.

However, the slow frequency is usually notched, so the native faster rhythm can be identified.

Slow alpha variant is differentiated from slow background of encephalopathy by:

- Notched appearance of the posterior rhythm with slow-alpha variant.
- Stereotypic appearance of the background with slow-alpha variant as opposed to polymorphic appearance of the posterior slow activity of encephalopathy.
- Normal central and frontal activity with slow alpha variant, as opposed to central, temporal, or frontal slowing in most cases of encephalopathy.

Encephalopathy rarely manifests as a slow background rhythm without any other sign of abnormality; the organization and frequency composition of the more anterior regions is key to differentiation.

Rhythmic Mid-Temporal Theta of Drowsiness

Rhythmic mid-temporal theta of drowsiness has been called *psychomotor variant*, but this latter term should be discarded, since it has diagnostic implications that are unsubstantiated. The rhythm consists of trains of sharply contoured waves in the theta range (Figure 4.6). They are most prominent in the temporal region but are also present in central regions. As the name suggests, the rhythm is seen mainly in drowsiness but may also be seen in relaxed wakefulness.

Rhythmic mid-temporal theta of drowsiness is differentiated from seizure activity by:

- Normal background before and after the rhythm
- Absence of a progressive frequency change that typifies seizure activity
- Presence in drowsiness but not in sleep

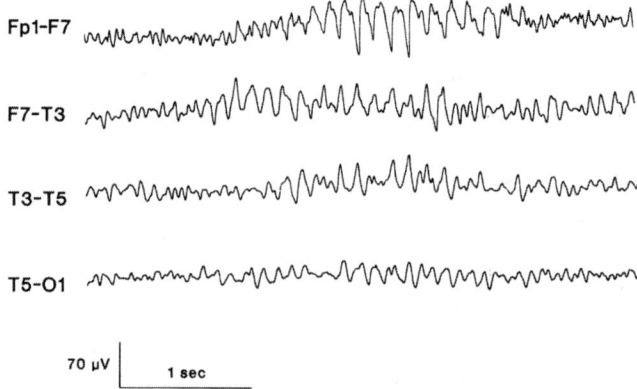

FIGURE 4.6 *Rhythmic temporal theta of drowsiness seen with maximum in the anterior temporal region. This is the left lateral portion of the longitudinal bipolar montage.*

This pattern is considered a normal variant, but has been described in association with structural lesions.

Subclinical Rhythmic Electrographic Discharge of Adults

Subclinical rhythmic electrographic discharge of adults (SREDA) is rhythmic sharp waves that evolve into a rhythmic theta pattern. It has an abrupt onset and changes in conformation and frequency throughout the discharge. SREDA is seen in older patients during the awake state. SREDA can easily be confused with seizure activity. Some important differentiating features include:

- Occurrence only in the waking state
- Intact consciousness during the discharge

Mittens

Mittens are seen only in sleep and consist of a partially fused sleep spindle and vertex wave. The last wave of the spindle is superimposed on the rising phase of the vertex wave. The voltage summation gives the last spindle wave a faster appearance that simulates a spike. The name is derived from the appearance of a mitten, the thumb being the fused spindle wave and the hand portion being the slow component of the vertex wave.

Mittens are normal, but can be confused with spike–wave patterns. Careful examination can usually dissect out the components constituting the mitten. While older literature correlated mittens to psychiatric disorders and tardive dyskinesia, this should still be considered a normal pattern.

Noncerebral Potentials

Eye Movement

Eye movement artifact is seen in the anterior leads in virtually all records. The eye is polarized with the cornea positive relative to the fundus. Therefore, when the eye rotates to look down, the leads over the frontal region are close to the negative end of the ocular dipole. This effect is most prominent for Fp1, Fp2, F3, and F4. The reverse is true with upward gaze, as the frontal leads are close to the positive end of the dipole. With lateral gaze, the electrodes most affected are F7 and F8. For example, with left gaze, F7 becomes more positive, while F8 becomes more negative.

Differentiating eye movement artifact from electrocerebral activity is usually not difficult. First, eye movements have a stereotypic pattern that looks different from most abnormal frontal slow activity, which is usually more polymorphic. The onset of the slow waves caused by eye movement is rapid with a slower decay. Also, eye movement waveforms are typically superimposed on a normal low-voltage, high-frequency background. Abnormal frontal delta activity is usually associated with increased theta activity and reduced beta activity in the frontal regions. If identification

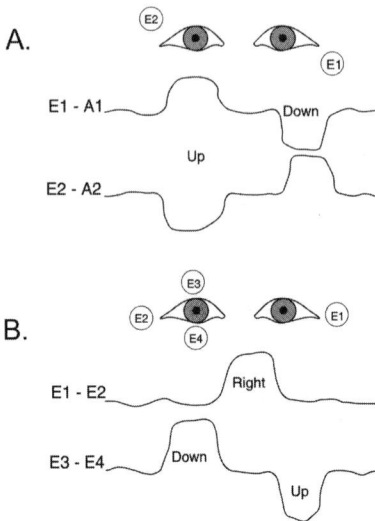

FIGURE 4.7 *Eye lead placement.* **A.** *Eye lead placement and sample recordings with vertical eye movements. This positioning helps with differentiation of eye movements from frontal slow activity.* **B.** *Alternative method of recording eye movements. This allows for differentiation of horizontal from vertical eye movements.*

is in doubt, eye leads should be placed to definitively distinguish between cerebral and eye activity.

Eye leads can be placed in several ways. The two most common methods are shown in Figure 4.7. We use the method shown in Figure 4.7A. Electrodes are placed above and lateral to the right eye and below and lateral to the left eye. These electrodes are referenced to an average or ear electrode. With upward gaze, the positive end of the dipole rotates toward the right lead but away from the left. This causes pen deflections of opposite polarity in the recording. With left gaze, the positive cornea rotates toward the left lead but away from the right. Again, the pen deflections will be in opposite directions. These electrode derivations will detect slow activity in the frontal lobes, but this slow activity will not reverse between the two sides. Therefore, in these channels, slow activity that is opposite in polarity is of ocular origin, while slow activity that is of the same polarity on both sides is most likely of cerebral origin.

The method shown in Figure 4.7 of eye movement detection allows for precise determination of direction of gaze. Vertical gaze can be distinguished from horizontal gaze. This is seldom of interest on routine EEG testing.

Eye Opening

Eye opening results in not only the eye movement artifact described above, with the eye moving up and down with the blink, but also alters the posterior rhythm. There is prompt attenuation of the posterior dominant rhythm with replacement with low-voltage fast activity.

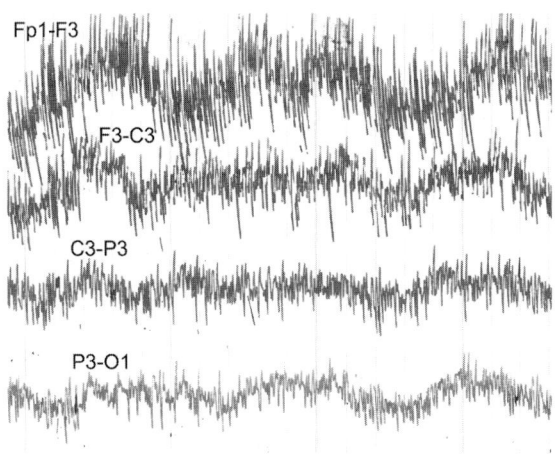

FIGURE 4.8 *Marked muscle artifact in a tense patient. This is the left parasaggital portion of the longitudinal bipolar montage.*

Eye closure produces eye movement artifact along with prompt restoration of the posterior dominant rhythm, often with a brief increase in frequency, referred to as *squeak*.

Muscle Artifact

EMG activity is frequent contaminant of EEG recordings. It is most prominent in the awake state and is characterized by fast, short-duration spikes in the temporal and frontal regions (Figure 4.8). Amplitude is approximately 50 µV. EMG activity is due to discharge of motor units in the temporalis and frontalis muscles. Distinction from spikes of cerebral origin is usually not difficult:

- EMG activity is very fast. In fact, the predominant frequency is much faster than the HFF setting of 70 Hz.
- EMG activity is not followed by a slow wave.
- EMG is most prominent in the waking, tense state, and disappears with relaxed wakefulness and sleep. In contrast, epileptiform activity is often best seen in drowsiness and sleep.
- EMG activity can often be attenuated by asking the patient to open his mouth. The masseter and pterygoid muscles do not contribute much to surface EMG activity, so this relaxation promotes decreased activation of scalp muscles and a reduction in scalp muscles without adding further artifact from jaw muscles.

Glossokinetic Artifact

The tongue is negative at the tip in comparison to the base, forming a dipole similar to the eye. Movement of the tongue is usually seen in the waking state as delta activity that can be mistaken for pathologic delta activity.

Talking can exacerbate the glossokinetic artifact. The artifact disappears with drowsiness and light sleep.

Differentiation of glossokinetic artifact from electrocerebral slow activity can be difficult and may require technician observation. It is usually concurrent with muscle artifact of the temporalis and frontalis. If there is still any doubt, electrodes should be placed below the orbits. These electrodes are too distant to detect electrocerebral activity but will pick up the slow activity due to tongue and eye movement. The patient can be asked to say "la la la" so that the potential field and appearance of glossokinetic artifact can be identified.

Movement Artifact

Movement produces artifact in two ways: movement of the electrode leads and perturbation of the electrode–gel interface.

Electrode Leads

The flow of electrons through wires creates a weak magnetic field. Magnetic fields can in turn influence the flow of electrons through conductors—they induce current. This interaction between current and magnetic fields is termed *inductance* and is the physical basis for inductors, discussed previously. Inductance is an important mediator of noise.

Electrode leads are unshielded and therefore susceptible to the effects of ambient magnetic fields. These magnetic fields are created by current in nearby power lines and equipment. Magnetic fields then induce current to flow through the electrode leads. This mechanism is termed *stray inductance* because the inductance is not intentional within the machinery. The induced current flows through the electrode leads just as signal current does. The amplifier cannot distinguish between signal and noise, so both are amplified. Since line power is 60 Hz, this creates 60 Hz interference.

The other major source of 60 Hz interference is *stray capacitance*. There is a small amount of capacitance between electrode leads and between these leads and power lines. A charge can build up across this capacitance. The capacitance will change with electrode lead movement, thereby altering the built-up potential. Also, alternating current in power lines can create a small capacitative current in electrode leads, thereby creating 60-Hz interference.

The differential amplifier will reject much of the 60-Hz interference; however, the rejection is incomplete under the following conditions:

- If electrodes are affected unequally by stray inductance and capacitance (because of the difference in lead position and proximity to electric wires)
- If there are unequal electrode impedances
- If there is electrode movement

When the electrodes are stable, stray inductance can cause 60-Hz interference that is rejected by the differential amplifiers. When there is electrode movement, however, the orientation of the electrode leads in space is altered. This produces a sudden change in induced current. The transient alternation in current is interpreted by the amplifier as a voltage shift.

Electrode–Gel Interface

A diffusion potential is established at the interface between the electrode and the gel that is caused by movement of ions between the two substances. The concept is similar to the creation of a resting membrane potential in neuronal membranes. When there is movement of the head, this junction is disturbed and the junction potential discharges, injecting current through the electrode into the input amplifier. This is interpreted by the amplifier as a voltage pulse.

The appearance of this artifact is a brief spike followed by a gradual decay to baseline. During the pulse, the responsiveness of the amplifier may be reduced, since the amount of current flow from the artifact can be sufficient to overload the input amplifier. If there is no further movement, a new equilibrium is established.

Machine Artifact

Machine artifact is usually a high-frequency artifact that may be at 60 Hz, the frequency of line power. This is true if the origin is a power supply, motor, or other AC-powered component. However, since most electronic circuits are DC-powered, machine artifact is more commonly due to DC-powered motors or resonant electric activity.

Motors are contained in IV pumps, ventilators, and hospital beds, among other devices. Figure 4.9 shows the machine artifact created by an air bed. Briefly unplugging the bed results in a hard surface for the patient, but a tremendous improvement in the EEG recording.

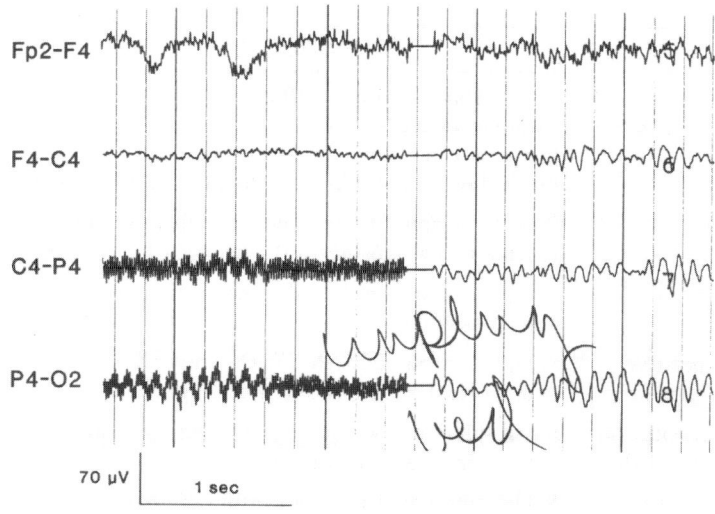

FIGURE 4.9   *Machine artifact. This is the normal EEG recorded with the right parasaggital portion of the longitudinal bipolar montage. The high-frequency artifact seen in the posterior channels, with most prominence in the P4 electrode, disappears when the electric bed is unplugged.*

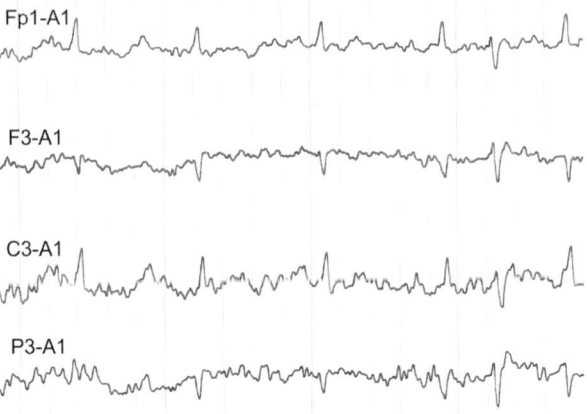

FIGURE 4.10 ECG artifact is seen in this left parasaggital portion of the ear reference montage. Ear reference montages are much more likely to show ECG artifact than bipolar or average reference montages.

Electrical artifact can be generated by equipment and perceived as machine noise. Among the devices with inherent resonance are radios.

Electrocardiogram Artifact

ECG is recordable from some patients, especially young children and with the use of referential montages (Figure 4.10). Increased interelectrode distance predisposes to increased ECG artifact. If there is a question as to whether there is ECG artifact, a separate channel with cardiac leads can be placed and the relative timing of the brain spikes and ECG compared.

Pulse Artifact

Pulse artifact is related to ECG and is due to movement of an electrode over an artery. Pulsation of the vessel moves the electrode and lead slightly, altering the diffusion potential of the electrode–gel connection and altering the electromagnetic interaction between adjacent electrode leads. This produces an irregular delta wave that can be localized to the lead overlying the artery. Identification of pulse artifact can be made by:

- Localizing the artifact usually to one electrode; cerebral structural lesions would produce slowing recorded by more than one electrode.
- Timing of the pulse to the ECG; there is a brief delay between the ECG QRS complex and the pulse but the timing should be consistent.
- Otherwise normal background; patients with polymorphic delta activity due to a structural lesion will typically have an abnormal background,

with theta activity and absence of a well-developed posterior dominant rhythm.

If there is still doubt, then examination of the scalp can show the pulsing vessel. Movement of the electrode off the vessel can confirm the artifact, though this is seldom necessary.

Activation Methods

Activation methods are used to bring out epileptiform discharges in patients with suspected seizure disorders. Hyperventilation and photic stimulation are routinely used. Sleep is considered by some to be an activation method, but is considered separately, above, since it is a normal state change rather than an applied method.

Photic Stimulation

Photic stimulation is repetitive flashes of light delivered at different rates. The flash stimulus can evoke epileptiform discharges at certain flash rates. Photic stimulation is more likely to evoke epileptiform discharges in patients with generalized epilepsies than in patients with focal epilepsies.

Methods
The stimulating protocols are preprogrammed into most EEG machines. General guidelines for performing photic stimulation include the following:

- Train duration of 10 seconds
- Trains delivered every 20 seconds
- Initial flash rate of 3/sec
- Higher rates for successive trains. We use 3, 5, 7, 9, 11, 13, 15, 18, 20, 24, and 30 flashes per second.

If the discharge is activated as a specified frequency, the technician should repeat that frequency at the completion of the photic stimulation routine.

Normal responses to photic stimulation include the visual-evoked response, the driving response, and the photomyoclonic response.

Normal Photic Response
Visual-evoked response: A visual-evoked response can be seen in occipital leads at flash frequencies of less than 7/sec. This response is the same as the flash-induced visual-evoked potential discussed in Chapter 14, but is much more variable because the response is not averaged (Figure 4.11).

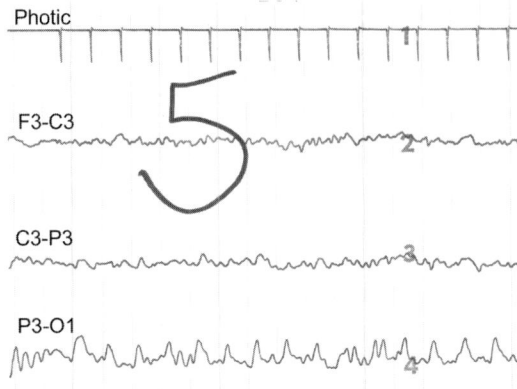

FIGURE 4.11 *Photic-evoked potential. Left medial portion of the longitudinal bipolar montage, with channel 1 showing the stimuli. Flash at 5/sec produces an evoked potential in the fourth channel, due to activity in the occipital lead. The upgoing potential in this bipolar montage indicates positivity at the O1 electrode. The positivity is delayed from the stimulus by about 100 msec, indicating that this is an evoked potential rather than a photic response.*

Driving response: A driving response is seen at flash frequencies of 7/sec and greater (Figure 4.12). The two responses look alike but are distinguished by their temporal relation to the stimulus. The visual-evoked response occurs approximately 100 ms after the stimulus, and the driving response is exactly time-locked to the stimulus.

The absence of a visual-evoked response or a driving response is not abnormal unless it is well developed on one side and absent on the other. Such asymmetry suggests an abnormality affecting either the projections from the lateral geniculate to the cortex or the calcarine cortex, itself.

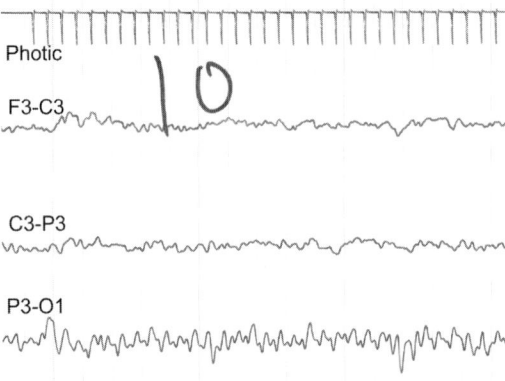

FIGURE 4.12 *Photic driving response. Photic stimulation at 10/sec produces a fast response that is time-locked to the stimulus, differentiating it from the evoked potential. Frequencies that produce a driving response are faster than those that produce an evoked response. Left medial portion of the LB montage.*

Photomyoclonic Response

The photomyoclonic response is caused by repeated contraction of frontal muscles that are time-locked to the flash stimulus with a delay of 50–60 ms. The potentials are suppressed by eye opening and disappear with neuromuscular blockade. Muscle activity that may have the appearance of seizure discharges appears in the anterior leads. Several factors help to distinguish between a photomyoclonic response and a photoconvulsive response:

- The photomyoclonic response is anterior, whereas the photoconvulsive responses are posterior or generalized.
- The photomyoclonic response stops promptly at the end of the stimulus train, whereas the photoconvulsive response typically outlasts the stimulus.
- The spikes that make up the photomyoclonic response are much faster than those of cerebral origin. The synchrony of muscle fiber discharges is much greater than that of neuronal discharges.
- The photomyoclonic response has the same frequency as the flash, whereas photoconvulsive discharges are often slower, in the range of 3/sec.

Photoconvulsive Response

The photoconvulsive response is characterized by spike–wave complexes during photic stimulation (Figure 4.13). The discharge is usually activated only by a few specific flash frequencies. It never begins with the first flash and usually ends before the flash ends. The correlation of a photoconvulsive discharge with seizures is greatest if the discharge continues after the end of the flash train.

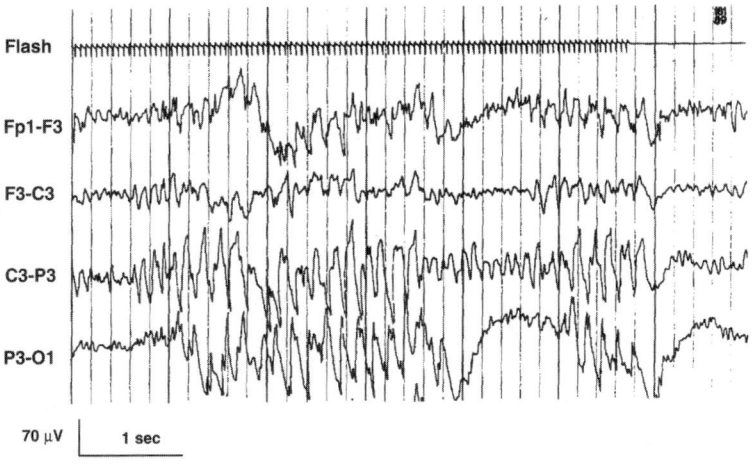

FIGURE 4.13 *Photoconvulsive response usually produces a specific range of photic frequencies. The discharge is not time-locked to the stimulus and typically lasts longer than the stimulus. Left medial portion of the LB montage.*

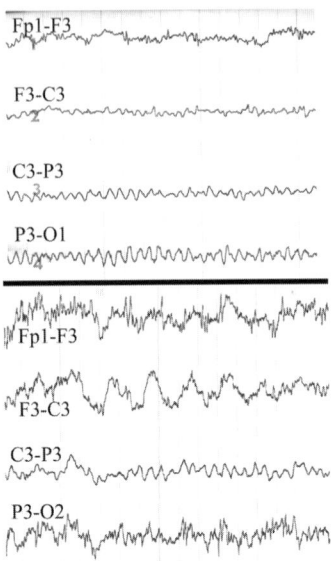

FIGURE 4.14 Hyperventilation. (Top) Normal waking EEG. Left medial portion of the LB montage. (Bottom) Slowing in the theta and delta range with hyperventilation. There is also muscle artifact.

Hyperventilation

Hyperventilation is usually used to activate the three-per-second spike and wave discharge of primary generalized epilepsy. In some patients, discharges are seen only during hyperventilation. The patient is asked to mouth-breathe deeply for approximately 3 minutes. If there is suspicion of absence seizures, the patient should hyperventilate for 5 minutes.

The normal response to hyperventilation is generalized slowing of the background activity in the theta range in both hemispheres (Figure 4.14). Absence of slowing is not abnormal and depends on effort, age, and time from last meal. Children show more slowing than adults with hyperventilation. Hypoglycemia may augment slowing.

Movement artifact may contaminate the record, especially in the posterior leads, as a result of head movement with chest excursions. Normal slow activity may have a notched appearance and should be interpreted as epileptiform. The epileptiform discharges activated by hyperventilation are usually not subtle.

Hyperventilation should not be performed in patients with cerebrovascular disease or intracranial hemorrhage. Hypocapnia and alkalosis may cause vasospasm and impair cerebral perfusion.

5 Abnormal Electroencephalographic Patterns

Definition of Abnormal

Abnormal can have two meanings. First, abnormal can mean that the finding is not expected in most patients. This says nothing about clinical implications. Second, abnormal can mean that the finding has clinical implications that the brain is not functioning the way it should. These are two very different meanings. For example, a sharply contoured slow wave might be considered to be normal, but is still unexpected. The term *normal variant* is used for EEG findings that are unexpected yet do not make the EEG interpreted as abnormal.

Slowing

Slow activity is frequently normal, such as theta during drowsiness and delta during sleep. However, focal delta during the waking state or theta for a posterior dominant rhythm in the waking state is clearly abnormal. Slowing can be divided into three classifications:

- Generalized slowing
- Regional slowing
- Focal slowing

Generalized slowing includes the slow activity that would characterize encephalopathy, with slowing of the posterior dominant rhythm, disorganization of the rhythm, and excessive theta activity anteriorly (Figure 5.1).

Regional slowing consists of slow activity that would only affect one portion of the brain yet not be focal to a single area. Examples indicating encephalopathy would be frontal intermittent rhythmic delta (FIRDA) or slowing of the posterior dominant rhythm. These two patterns are often associated with slowing outside of the regions, but the background may otherwise be normal.

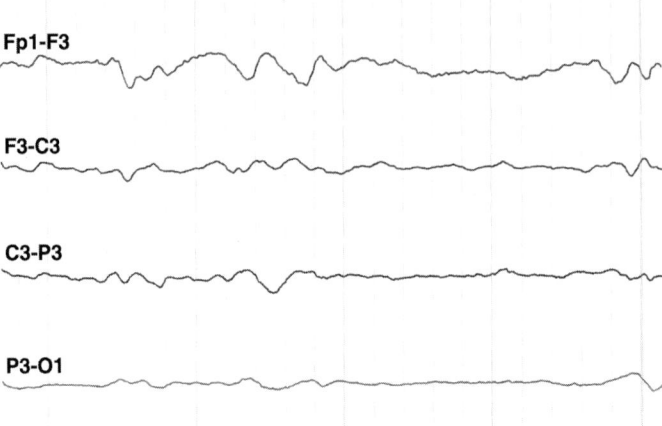

FIGURE 5.1 Generalized slowing. Left medial portion of the LB montage. The slow activity in the theta and delta range is inconsistent with a normal waking EEG in an adult.

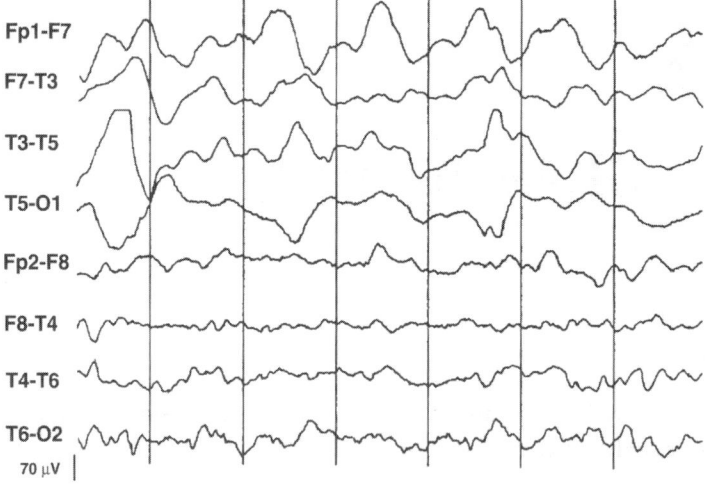

FIGURE 5.2 Focal slowing. Left and right portions of the LB montage showing generalized slowing, but more prominent high-amplitude slow activity from the top channels, left side of the head. This is focal slow activity superimposed on the generalized slow activity.

Focal slowing is usually indicative of a structural lesion, and includes focal theta activity and polymorphic delta activity (Figure 5.2).

Spike and Sharp Waves

Spikes and sharp waves are fast transients. The term *transient* means that the activity is episodic, and therefore stands out from the background. Spikes are usually surface negative and have a duration of 20–70 ms (Figure 5.3).

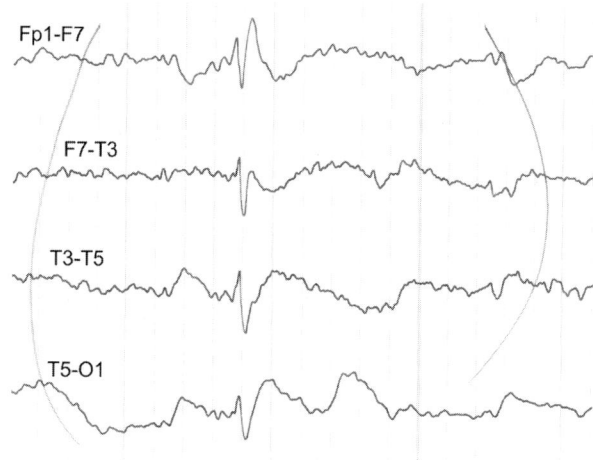

FIGURE 5.3 *Focal spike with maximal negativity at F7, the left anterior temporal region.*

TABLE 5.1   Differentiation between spike and nonspike potentials

Feature	Spikes	Nonspike potentials
Consistency of appearance	Stereotypes	Vary in morphology
Relation to the background	Stand out from the background	Embedded in the background
Rising phase	Rising phase faster than any other component	Rising phase may be slower than falling phase
Following slow wave	Has a following slow wave	Usually not followed by a slow wave
Effect of sleep	Activated by sleep	No change or disappearance with sleep

Sharp waves are also usually surface negative but have a duration of 70–200 ms. Potentials of less than 20 ms duration are usually not of cerebral origin, being either muscle fibers or electrical artifact. The negative pole of spikes and sharp waves is distributed across a region of the cortex, and this is the *potential field*. Electrical potentials are dipoles, so for each negative pole there is a positive pole. Since the scalp potentials are usually surface negative, the positive end of the dipole is usually subcortical and cannot be seen by scalp electrodes. However, occasionally the positive end of the dipole can be seen over a separate part of the cerebral cortex.

Table 5.1 lists criteria for differentiating spikes from nonspike potentials. Table 5.2 shows some clinical correlations with common disorders manifesting spikes. Among the major pitfalls is the possibility of misinterpreting sharply contoured rhythms as sharp waves and spikes. Theta activity can be pointed and give the appearance of spike potentials. A run of theta activity can look like a run of spike and wave complexes, especially if the theta is

TABLE 5.2 Electroencephalographic patterns with selected disorders that demonstrate spikes

Disorder	Electroencephalographic pattern
Rolandic epilepsy	Stereotyped spikes, often with triphasic appearance, followed by a slow wave Independent discharges with maxima near C3 and C4 Activated by sleep Background is otherwise normal
Occipital epilepsy	Unilateral or bilateral high-amplitude occipital spikes that are increased by light sleep and suppressed by eye opening Rapid discharge from the same regions during a seizure
Absence epilepsy	Classic three-per-second spike–wave or polyspike–wave complex Increased by hyperventilation Less well organized in sleep
Subacute sclerosing panencephalitis	Periodic high-amplitude bursts of sharp waves, often with an irregular delta wave superimposed Synchronous between hemispheres Duration of 0.5–2.0 sec Many seconds to a minute or more between bursts
Creutzfeldt–Jakob	Periodic complexes with a frequency of 0.5–2.0/sec Maximum anteriorly Abnormal background, with low-voltage slowing Periodic complexes abate during sleep
Anoxic encephalopathy	From normal to isoelectric depending on severity Common patterns include burst suppression and a periodic pattern that resembles CJD Alpha coma is a diffuse nonreactive alpha activity with a frontocentral predominance; these latter patterns indicate a poor prognosis
Herpes simplex encephalitis	Generalized slowing with a frontotemporal predominance Periodic complex of sharp waves or sharply contoured slow waves, that appear irregularly High-amplitude discharges with a frequency of 0.2–1.0/sec
Arbovirus encephalitis	Slowing in the theta and delta range with little intra- or interhemispheric synchrony Few faster frequencies
Lennox–Gastaut syndrome	Slow spike–wave complex Slow background in many patients Increasing disorganization during sleep
Juvenile myoclonic epilepsy	Generalized polyspike discharges followed by a slow wave Higher amplitude in the frontal region Otherwise normal background
Complex partial seizures	Various patterns depending on site of origin of epileptiform activity May be unilateral temporal or frontal spikes Focal slowing or no discharge recordable with surface electrodes
Simple partial seizures	Midline spikes with a prominent negative phase or biphasic in most patients Occasional patients may have a positive prominence on surface EEG These patterns correlate with simple motor seizures
Generalized tonic–clonic seizures	Generalized polyspike discharge that often, but not invariably, has a slow wave Ictal activity is often high-frequency spikes without obvious slow waves

notched or otherwise sharply contoured. A sharp wave should not be considered abnormal unless it stands out from the background and is reproducible. The alpha rhythm may occasionally appear sharp, even though the pointed component is positive. Positive polarity helps to distinguish this waveform from a spike potential.

When Spikes and Sharp Waves Are Normal
Most spikes and sharp waves in adults are abnormal unless they are artifacts. Several normal spike-like potentials can be seen, however. These include:

- Vertex waves
- Occipital lambda waves
- 14- and 6-Hz positive spikes
- Wicket spikes
- Benign epileptiform transients of sleep (BETS)
- Positive occipital sharp transients of sleep (POSTS)
- Six-per-second (phantom) spike and wave

These are discussed in detail in Chapter 4, and can sometimes be difficult to distinguish from epileptiform activity. The relevant discussions should help with this distinction.

Spikes and Sharp Waves Not Associated with Seizures
The electrophysiologist is not blind to the clinical question when interpreting the EEG. The best approach is to describe the record, a phase of interpretation that would be independent of the clinical question. The impression will then consider the EEG findings in light of the clinical question. Therefore, detailed clinical information provided to the electrophysiologist certainly aids clinical utility of the EEG. Most of the time that true spikes and sharp waves are seen, the clinical correlation is seizures, and in this case, the impression might read:

> "Abnormal study because of epileptiform activity arising from the right temporal region. This would be consistent with a partial seizure disorder with a focus in the right temporal lobe."

However, if the clinical question is not seizures, the clinical interpretation of these findings is more controversial. We believe that mention should be made of the epileptiform appearance of the activity but also the disclaimer made that not all patients with epileptiform activity have seizures. For example, the same EEG pattern discussed above in a child with behavioral disorder might result in the following impression:

> "Abnormal study because of epileptiform activity arising from the right temporal lobe. In the appropriate clinical situation, this would be supportive of a partial seizure disorder, however, the presence of this pattern of epileptiform activity does not mean that the patient is certain to have seizures."

Focal vs. Generalized Abnormalities

Abnormal patterns can be focal or generalized, and the clinical implications depend on the exact pattern and location. However, there are some generalizations:

- *Focal slowing*—usually suggests a focal structural lesion underlying the scalp electrodes.
- *Focal spikes or sharp waves*—can correlate with a focal structural lesion but more commonly suggests a partial seizure disorder. The type of seizure correlates with location. Of course, not all sharp activity indicates seizures.
- *Diffuse slowing*—usually associated with encephalopathy, which can have many potential causes, including toxic, metabolic, degenerative, and multifocal vascular disease.
- *Diffuse spikes or sharp waves*—correlate with a generalized seizure disorder. Cannot rule out secondary generalization.

Abnormal Frequency Composition

Interpretation of the EEG is a complex assessment of frequency composition and localization. Abnormal frequency composition can consist of any of the following:

- Slowing of the background rhythms
- Excessive fast activity
- Excessive theta activity

Slowing of the background rhythms is discussed below in "Slow Activity." Excessive fast activity is usually seen in patients sedated with benzodiazepines—beta activity is prominent frontally. Theta activity is present in almost all recordings, and can be seen if the gain is high enough. Theta is not a prominent component of the background in waking adults, and when it stands out from the baseline is abnormal; there are several potential clinical correlations, discussed below.

Slow Activity

Diffuse Slowing

Diffuse slowing can have several presentations. The most common is slowing of the posterior dominant rhythm in the waking state. Occasionally, slow activity can be superimposed on an otherwise normal waking background. Identification of abnormal slow activity in sleeping records is especially challenging.

Slowing of the Posterior Dominant Rhythm

The normal adult waking EEG consists of mainly fast rhythms. With eyes closed, rhythms in the alpha range are seen from the posterior regions and faster frequencies are seen from the frontal regions. Slowing of the posterior dominant rhythm (PDR) to less than 8.5 Hz is always abnormal in adults. Slowing of the PDR is usually seen as the posterior rhythm in the theta range, for instance 6–7 Hz. The slow PDR differs from the normal faster rhythm in a few ways:

- Slow PDR is less stereotyped than normal PDR, with bumps on the waves.
- Slow PDR is less reactive to eye opening than normal PDR, it does not show the degree of attenuation of normal PDR.
- Slow PDR is often associated with theta prominent more forward of the occipital regions than the normal PDR extending forward of the occipital regions.

Interpretation of the Slow Posterior Dominant Rhythm The slow PDR is interpreted as being abnormal, but is not specific. Possible causes include:

- Toxic–metabolic encephalopathy
- Degenerative dementia
- Multifocal vascular disease

The impression when this is the only finding might be: "Abnormal study because of slowing of the posterior dominant rhythm. This is suggestive of a diffuse encephalopathy, although it is a nonspecific finding." Comment might be made about metabolic, toxic, and dementing causes, depending on the specified clinical question.

Subharmonic Posterior Dominant Rhythm (Normal Variant) Occasionally, the PDR has the appearance of a subharmonic—where there may appear to be a 5–6-Hz PDR with otherwise normal frequency composition and appearance of the EEG. This is a normal variant, and should be interpreted as normal. The subharmonic PDR can be differentiated from slowing of the PDR in the following ways:

- Slowing of the PDR in the 5–6-Hz range should be associated with slowing seen anteriorly to the occipital lobes, whereas subharmonic PDR has otherwise normal frequency compositions.
- Slowing of the PDR in the 5–6-Hz range will usually not attenuate completely to eye opening, whereas subharmonic PDR completely attenuates.
- Slowing of the PDR in the 5–6-Hz range will have an irregular, polymorphic appearance, whereas subharmonic PDR is regular, and usually notched, so that the underlying 10-Hz rhythm can be seen.

Slow Activity Superimposed on the Waking Background

Theta and delta activity in waking records is usually abnormal. Diffuse slowing is usually polymorphic delta or irregular theta that is seen from both hemispheres. The slowing does not typically have the regional concentration of the normal PDR or frontal fast activity. Most of the time, the PDR is slow when there is diffuse slowing, but not always. Diffuse theta with a temporal prominence can be associated with a PDR still in the alpha range.

Causes of slow activity superimposed on an otherwise normal waking background include:

- Encephalopathy due to toxic or metabolic causes
- Cerebrovascular disease that is multifocal or diffuse
- Head injury

Generalized Slowing in Sleep Recordings

Identification of abnormal slowing in a waking record is easy, but identification of abnormal slowing in a sleeping record is much more difficult. The sleep record consists of slow activity in the theta and delta range, and the exact pattern depends on sleep stage. For example, stage 3 and 4 sleep is composed of prominent delta activity with virtual abolition of the faster frequencies. These deeper stages of sleep could easily be misinterpreted as encephalopathy if the electrophysiologist is more used to seeing waking records and stages 1 and 2 of sleep. Therefore, encephalopathy should be the interpretation of a sleep record only if the slow activity is inconsistent with any stage of the sleep-wave cycle. Even then, a waking record should be examined, if at all possible.

Conversely, a normal sleeping record does not rule out an encephalopathy. It is very common to have abnormal slowing during the wake state yet normal sleeping patterns. In this instance, the EEG report should reflect the limitations on interpretation of encephalopathy in the sleeping state. The impression might read:

> *Normal sleeping EEG. Encephalopathy is difficult to diagnose in the sleeping state. A waking record should be obtained, if clinically indicated.*

Focal Slowing and Polymorphic Delta Activity

Focal slowing usually indicates a focal structural lesion of the hemispheres. Structural lesions of the brainstem produce generalized slowing, if any abnormality at all. Focal slowing usually overlies the lesion, but the correlation is not exact.

Focal slowing is irregular and composed of delta activity with theta activity superimposed. Even faster activity is then superimposed on this background. This irregular appearance is the reason for the term *polymorphic delta activity* (PDA).

PDA may not be continuous, although this is surprising when considering the pathologic processes that cause PDA and the physiology behind the EEG patterns. Why should a fixed structural lesion create a synchronous potential shift that would be episodic? The PDA often appears on a disorganized EEG background, but the background may actually be normal.

The neurophysiological substrate of PDA is not completely understood. In general, PDA is interpreted as being an abnormality in the white matter relays between the cortex and subcortical nuclei.

PDA is the most common finding in focal structural lesions such as tumors, contusion, hemorrhage, infarction, and abscess. The presence of focal spikes or sharp waves without another disturbance on the background is seldom a sign of a focal parenchymal lesion. Focal slowing is nonspecific; there are no characteristics that distinguish one cause from another. Complicated migraine and postictal state may cause focal slowing.

Intermittent Rhythmic Delta Activity

Intermittent rhythmic delta activity is always a sign of cerebral dysfunction and is thought to be due to a disconnection of the cerebral cortex from the deep nuclei. Slow activity is seen at about 2.5 Hz with a distribution that depends on age and pathology. In adults, the rhythmic slow activity is usually frontal, hence *frontal intermittent rhythmic delta activity* (FIRDA, Figure 5.4). In children, the slowing is commonly seen in the occipital regions, hence the term *occipital intermittent rhythmic delta activity* (OIRDA); this acronym is hard to pronounce, so posterior is substituted for occipital, hence, *PIRDA*.

The rhythmic slowing of FIRDA and PIRDA may last for several seconds then disappear for longer intervals, hence the intermittent nature of the rhythm. The slow activity is augmented by eye closure or hyperventilation, but attenuated by stimulation or by non-REM sleep. FIRDA reappears in REM sleep.

A wide variety of lesions can produce IRDA, so the interpretation should indicate the abnormal nature of the rhythm without implications for localization. There are no major diagnostic differences between FIRDA and PIRDA. PIRDA is seen occasionally in children with absence epilepsy. Both FIRDA

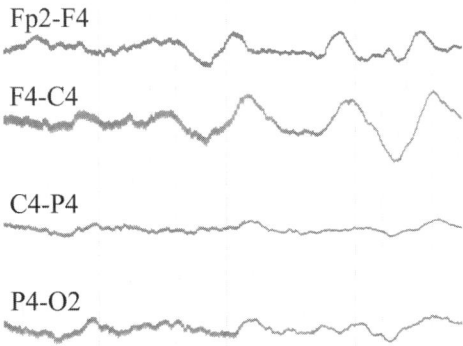

FIGURE 5.4 *Frontal intermittent rhythmic delta activity (FIRDA), shown on the left medial portion of the LB montage.*

and PIRDA can be caused by:

- Midline tumors
- Metabolic encephalopathy
- Degenerative disorders
- Some encephalitides

FIRDA is differentiated from PDA by the latter's lack of reactivity to the stimulus, usual unilateral appearance, lack of rhythmicity, and the continuous appearance.

Slow Activity as a Seizure Discharge

Seizures occasionally manifest on routine EEG as rhythmic slow waves. Presumably, the spike component is either very small in amplitude or not projected to the cortical surface. Differentiating epileptiform slow waves from FIRDA, PIRDA, and PDA can be difficult. Epileptiform slow activity interferes with the normal background, whereas FIRDA may be associated with an otherwise near-normal background. Epileptiform slow activity is differentiated from PDA by the stereotypic nature of the epileptiform activity. Epileptiform waves tend to be smoother, and if the discharges are bilateral, there is usually a high degree of interhemispheric synchrony.

Focal Loss of Electroencephalography Patterns

Focal attenuation of EEG activity usually indicates a structural lesion. Beta activity is most sensitive to this effect. Occipital lesions can cause unilateral loss of the posterior alpha. Unilateral lesions may also disrupt sleep patterns so that sleep spindles, vertex waves, or both are not seen from the affected hemisphere.

Unilateral suppression is commonly seen with subdural hematoma. Caution is required in these cases in that the background from the opposite hemisphere may be slow due to either trauma or midline shift with compression. In this situation, the neurophysiologist may focus attention on the side of higher amplitude with prominent slow activity and erroneously interpret the affected side as desynchronized.

Spikes and Sharp Waves

Focal Sharp Activity

Focal spikes and sharp waves usually indicate a seizure disorder of with partial onset, but focal discharges can also be indicative of a structural lesion in the absence of seizure activity. Frontocentral discharges may be seen in patients with simple partial seizures. Temporal or frontal spikes may be seen in patients with complex partial seizures. Normal focal spike–wave complexes include 14- and 6-Hz positive spikes, subclinical rhythmic electrographic discharge of adults (SREDA), and wicket spikes.

Focal spikes are interpreted if the spike is consistent, has an identifiable field, and cannot be explained by artifact. A single spike during the course of a recording should not be interpreted as abnormal, although if the conformation is worrisome, a note in the body of the report should be made. Also, great caution should be exercised when interpreting a spike that is seen from a single electrode; remember that a single electrode is usually represented on more than one channel in a montage. Table 5.3 shows guides for interpretation of focal spikes.

Focal Spikes Associated with Seizures
Focal spikes are associated with partial seizures and the benign epilepsies of childhood. Partial seizures are divided into simple and complex, based on symptomatology rather than EEG findings. The benign epilepsies of childhood can manifest as focal and generalized seizures.

Simple Partial Seizure The EEG during a simple partial seizure usually shows prominent spiking over the involved cortex, although in some patients there may be localized slowing that may become generalized. A typical pattern might be left central spikes in a patient who presents with focal seizures affecting the right arm. Occasionally, the sharp component of the discharge may be subtle or missing. The epileptiform activity may occur in deep layers of cortex and subcortical structures so that the spike

TABLE 5.3 Interpretation of focal spikes

Type	Electroencephalographic features	Clinical features
Rolandic spikes	Spike and slow-wave complex Often triphasic and fast Maximal at C3 and C4	Benign rolandic epilepsy May be seen in subjects without seizures
Occipital spikes	Negative or biphasic spikes over the occipital region Unilateral or independent bilateral discharges	May be benign occipital epilepsy, but occipital spikes are not always benign May be seen in blindness
Parietal sharp waves	Sharp waves or spikes in the parietal region Can be activated by forehead taps	Often associated with versive head and eye movement or sensory seizures
Temporal sharp waves	Sharp waves in the temporal region Anterior temporal = F7, F8 Mid-temporal = T3, T4 Posterior temporal = T5, T6	Anterior temporal sharp waves often associated with partial complex seizures Mid-temporal sharp waves associated with seizures and with psychological complaints Posterior temporal sharp waves associated with seizures in the majority of patients, often generalized tonic–clonic Also associated with psychological complaints
Periodic lateralized epileptiform discharges	Unilateral or bilateral independent sharp and slow-wave complexes at 1–2/sec	Any destructive process Often anoxia, herpes simplex encephalitis, stroke, tumor

potentials are not projected to the surface electrodes. Alternatively, there may not be sufficient synchrony to produce a spike detectable on the surface.

Partial seizures may spread throughout the hemispheres, resulting in a generalized seizure. This is secondary generalization, as opposed to primary generalized seizures, like absence. Secondary generalized seizures may have a focal onset that can be detected clinically, but this is not always the case. The generalization may occur so quickly that the focal onset can only be determined by EEG, and not by clinical appearance.

Complex Partial Seizure The EEG during complex partial seizures usually shows focal spikes in the temporal or frontal region. Routine EEG may not detect the spikes if they originate in cortex that is not directly underlying the surface electrodes. Sphenoidal, nasopharyngeal, or depth electrodes may be needed to identify these discharges.

Complex partial seizures may have secondary generalization. If the rate of generalization is fast, the partial origin may not be evident, and the only clue would be separate complex partial seizures without generalization and EEG showing focal activity prior to the generalization.

Benign Focal Epilepsies of Childhood Benign focal epilepsies of childhood are so called because they are age-related and seldom persist into adult life. There are two types: rolandic and occipital.

Rolandic Epilepsy Rolandic epilepsy is characterized by interictal discharges arising from the central regions, localized near electrodes C3 and C4 (Figure 5.5). The interictal discharges are independent and augmented by sleep. Relatives of patients with rolandic epilepsy may have EEG abnormality as a genetic marker without clinical seizures.

The discharges of rolandic epilepsy are so characteristic in location and pattern that they are seldom confused with other pathologic activity.

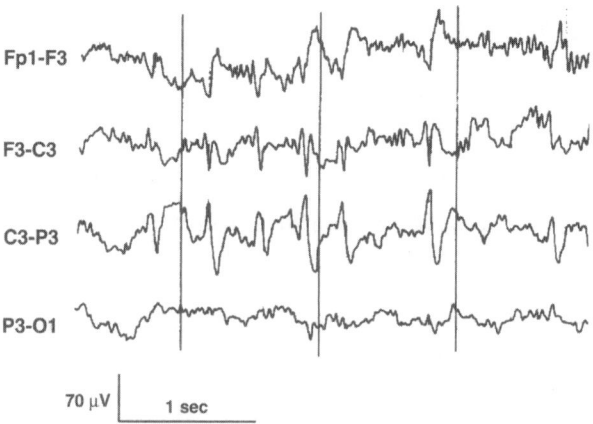

FIGURE 5.5 *Central spikes, of the types seen in rolandic epilepsy. The maximal negativity is at C3. Discharges are expected on the right side, as well, although the right hemisphere channels are not shown in this figure.*

Independent central spikes are seen on an otherwise normal background. This must be differentiated from multifocal spikes, however.

Occipital Epilepsy   Occipital epilepsy is characterized by interictal sharp waves with predominance at O1 and O2. Rolandic and occipital epilepsy may occur in the same families, and relatives with no history of seizures may have either occipital or rolandic discharges on EEG.

During the seizure, the EEG shows 2–3/sec spike–wave discharges with predominance in the occipital region. The interictal discharge may be blocked by photic stimulation or eye opening.

Focal Sharp Activity Without Seizures

Focal spikes or sharp waves are occasionally seen in patients with no clinical seizures. The EEG may have been ordered for some nonepileptic indication. Some of these are children who are genetic carriers of benign epilepsies, whereas in others there is no explanation. The interpretation of these records is controversial. Some neurophysiologists believe that all abnormal sharp activity is potentially epileptogenic and should be interpreted as such. Unfortunately, this may result in unneeded use of antiepileptic drugs. Patients should be treated with antiepileptic drugs based on clinical presentation rather than on EEG findings. The old adage still is valid "Treat the patient, not the EEG." Of course, the countering argument is that the patient may have seizures that are not always clinically identifiable.

About 3% of normal individuals exhibit some sort of sharp activity, and in a much smaller percentage the appearance may be epileptiform, despite the absence of clinical seizures. The proportion is somewhat higher in children than in adults. Approximately 25% of these discharges are focal. Some of these patients will go on to develop seizures; however, these patients should not be treated with anticonvulsants without clinical evidence of convulsive activity. Of patients with seizures, about 50% will show abnormalities on EEG, but this value differs dramatically, depending on the clinical setting. Patients with absence epilepsy are more likely to have abnormal EEG than patients with complex partial seizures, for example.

Children with behavioral disturbances have been reported to have an increased incidence of focal sharp waves and spikes. The implication of these findings is controversial. Some investigators believe that the spikes may have contributed to the behavioral disturbance by interfering with normal social and intellectual development. Others believe that the spikes are incidental and should not be treated. The spikes are probably a reflection of brain dysfunction, which correlates with the behavioral disorder rather than being the cause of the dysfunction.

SREDA is sharply contoured rhythmic delta activity with prominence in the centroparietal region. This pattern is seen in older patients and has no definite clinical correlate. This is not an ictal discharge. Patients with this finding are said to be at increased risk for cerebrovascular disease, but the association is not convincing. This rhythm is not found in normal younger individuals and may be an abnormal pattern. The report should reflect the nonspecific clinical implications.

Some patients with congenital blindness may exhibit occipital spikes. These should not be interpreted as epileptiform.

Generalized Sharp Activity

Generalized sharp activity is usually associated with a slow wave, forming the spike–wave complex. Spike–wave complexes correlate better than single spike discharges with clinical seizures. Table 5.4 shows clinical correlations to generalized discharges. Generalized spike–wave discharges fall into four categories:

- Three-per-second spike–wave complex
- Slow spike–wave complex
- Fast spike–wave complex
- Six-per-second spike–wave complex

Three-per-Second Spike–Wave

The three-per-second spike–wave complex is usually equated with absence epilepsy. Although there is a strong correlation between the two, a patient with a three-per-second spike–wave complex may exhibit other seizure types, including generalized tonic–clonic seizures. The interpretation of such records should read: "This is an abnormal study because of three-per-second spike–wave complexes. This is consistent with a seizure disorder of the generalized type."

Appearance The three-per-second spike–wave complex is synchronous from the two hemispheres, with highest amplitude over the midline frontal region (Figure 5.6). The lowest amplitudes are in the temporal and occipital regions. The frequency changes slightly during the course of the discharge,

TABLE 5.4 Interpretation of generalized spikes

Type	EEG features	Clinical features
Burst suppression	Bursts of slow waves with superimposed sharp waves, interspersed on periods of relative flattening	Severe encephalopathy from anesthesia, anoxia, or other diffuse cause
Three-per-second spike–wave	Spike or polyspike complexes at 2.5–4.0/sec Increased by hyperventilation	Generalized epilepsy May be seen in clinically unaffected relatives
Six-per-second (phantom) spike–wave	Small spike–wave complexes May have a frontal or occipital predominance	If frontal, associated with tonic–clonic seizures If occipital, usually not associated with seizures
Hypsarrhythmia	High-voltage bursts of theta and delta with multifocal sharp waves superimposed, interspersed by relative suppression	Seen in children with infantile spasms

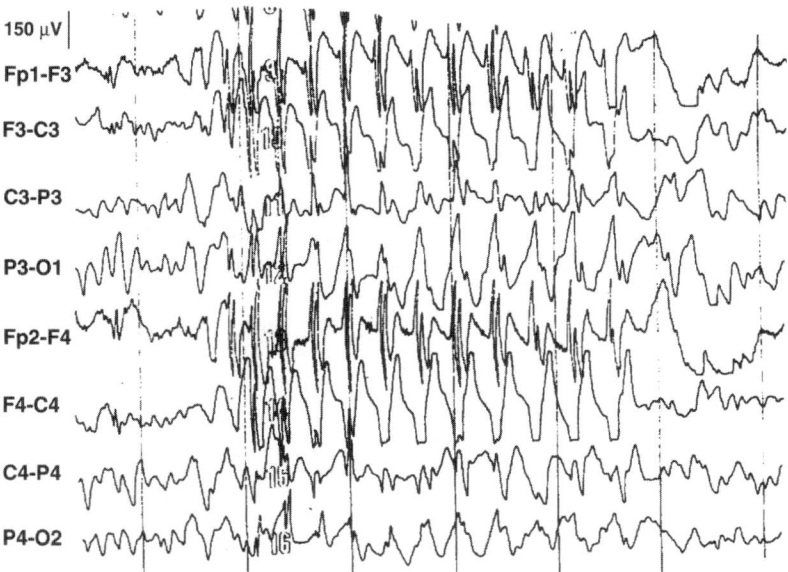

FIGURE 5.6 *The three-per-second spike–wave complex that is typical of absence seizures. The spike is actually a polyspike, and the frequency is not constant throughout the discharge, with faster frequency during the early portion of the discharge and slowing somewhat later in the discharge. Left and right medial portions of the LB montage.*

beginning close to 4/sec and declining to 2.5/sec. Immediately following the discharge, the record quickly returns to normal. The spike component may have a double spike or polyspike appearance.

The three-per-second spike–wave complex is promoted by hyperventilation. If absence epilepsy is considered, the patient should be asked to hyperventilate for 5 minutes instead of the usual 3 minutes. Children with absence seizures become symptomatic if the discharge lasts longer than 5 seconds. During the discharge, the technician should ask the patient a question. The patient with absence seizures often answers after the discharge. The question and the response should be noted on the record.

The three-per-second discharge is less well organized during sleep than during the waking state. Its appearance is more polyspike in configuration and the spike–wave interval is less regular.

The spike component is polyspike in some patients. Patients with this polyspike pattern are more likely to exhibit myoclonus.

Clinical Correlations The three-per-second spike–wave pattern correlates well with primary generalized epilepsy, if the remainder of the recording is normal. Factors that would make the clinician doubt the diagnosis of primary generalized epilepsy include:

- Abnormal EEG background
- Focal discharges

- History of slow neurologic development
- Abnormal neurologic examination

Treatment of absence epilepsy often abolishes the interictal discharge. This is different from most focal epilepsies in which interictal spiking persists despite good seizure control.

Slow Spike–Wave Complex

The slow spike–wave complex is at 2.5/sec or less. The morphology is less stereotyped than the three-per-second spike–wave complex. The duration of the slow spike is usually more than 70 ms, which is technically a sharp wave. The complex is generalized and synchronous across both hemispheres, with the highest amplitude in the midline frontal region.

During sleep, the slow spike–wave complex may be continuous. This activity may not indicate status epilepticus but rather represents activation of the interictal activity with sleep.

The slow spike–wave complex is frequently associated with the *Lennox–Gastaut syndrome*. It also has been called *petit mal variant*, but this term is misleading and should not be used. In the Lennox–Gastaut syndrome, the slow spike–wave complex is usually an interictal pattern, but may also be ictal. Since these patients have a mixed seizure disorder, ictal events may show patterns other than the slow spike–wave complex. Atonic seizures are characterized by generalized spikes during the myoclonus followed by the slow spike–wave pattern during the atonic phase. Atonic seizures are most characteristic of the Lennox–Gastaut syndrome. Akinetic seizures are characterized by the slow spike–wave discharge throughout the seizure. Tonic seizures occur in Lennox–Gastaut syndrome and are characterized by a rapid spike activity or desynchronization rather than the slow spike–wave complex.

Fast Spike–Wave Complex

The fast spike–wave complex has a frequency of 4–5/sec and has the appearance of slow waves with superimposed sharp activity, rather than distinct spike–wave complexes (Figure 5.7). Maximal amplitude is in the frontocentral region.

Patients have generalized tonic–clonic seizures with or without myoclonus. Absence seizures are rare. This is the most common pattern seen in patients with idiopathic generalized tonic–clonic seizures. The discharge is not as stereotyped as the three-per-second spike–wave complex, and the synchrony is less prominent.

Six-per-Second (Phantom) Spike–Wave Complex

The six-per-second spike–wave complex is characterized by brief trains of small spike–wave complexes that are distributed diffusely over both hemispheres, with a frontal or occipital predominance. They are most common during the waking and drowsy states and disappear during sleep.

The clinical implications of the frontal and occipital rhythms differ. Frontal predominance is frequently associated with generalized tonic–clonic seizures, whereas occipital predominance is not associated with clinical seizures.

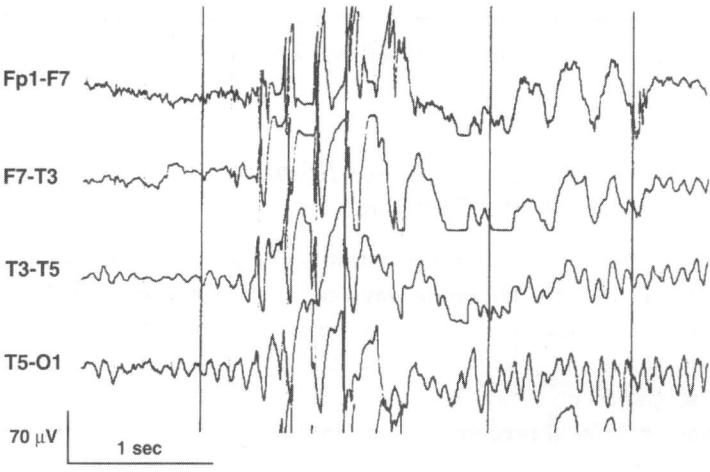

FIGURE 5.7 *The fast (about six-per-second) spike–wave complex that is seen in primary generalized epilepsies. Left lateral portion of the LB montage.*

Hughes (1980) provided the acronyms WHAM and FOLD:

- WHAM = waking record, high amplitude, anterior, males
- FOLD = females, occipital, low amplitude, drowsy

WHAM is associated with seizures and FOLD is not.

The six-per-second spike–wave pattern is differentiated from the 14-and-6 positive spike pattern not only by polarity but also by the more widespread distribution and occurrence in wakefulness. Both rhythms may appear in the same patient. The six-per-second pattern is interpreted as abnormal and the different clinical implications should be emphasized in the report.

Hypsarrhythmia

Hypsarrhythmia is characterized by high-voltage bursts of theta and delta waves with multifocal sharp waves superimposed (Figure 5.8). The bursts are separated by periods of relative suppression. In some circumstances, flattening of the EEG may be an ictal sign, indicating that there has been sudden desynchronization of the record.

Periodic Patterns

Periodic discharges usually indicate cortical damage, and can be due to stroke, anoxia, infection, degenerative disorders, and other conditions. The periodic patterns can be focal, regional, or generalized, with regional distribution being the most common.

Periodic Lateralized Epileptiform Discharges

Periodic lateralized epileptiform discharges (PLEDs) are high-amplitude sharp waves that recur at a rate of 0.5–3.0/sec (Figure 5.9). They are prominent over

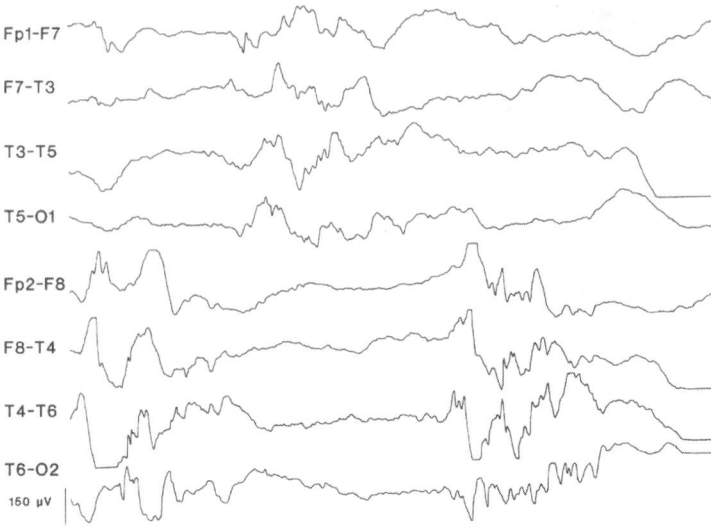

FIGURE 5.8 *Hypsarrhythmia, seen typically in children with infantile spasms. The high-amplitude bursts with interburst interval is characteristic. Left and right lateral portion of the LB montage.*

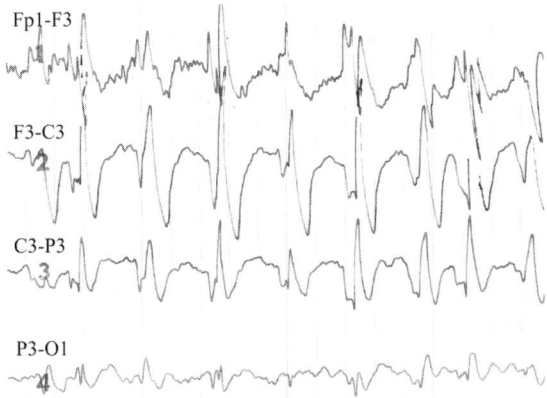

FIGURE 5.9 *Periodic lateralized epileptiform discharges (PLEDs) in a patient with herpes simplex encephalitis. Left medial portion of the LB montage.*

one hemisphere or one region. When bilateral, they are independent, thereby keeping the term *lateralized*. The background between the discharges is typically slow and suppressed.

PLEDs are a sign of parenchymal destruction and most commonly seen in strokes. Other important causes include head injury, abscess, encephalitis, hypoxic encephalopathy, brain tumors, and other focal cerebral lesions. It is impossible to distinguish definitively between causes on the basis of waveform. Of the encephalitides, herpes simplex most commonly produces PLEDs. Other viral infections produce slowing without PLEDs.

The PLEDs have an amplitude of 100–300 µV. An early negative component is followed by a positive wave. The discharge may be complex, with additional sharp and slow components superimposed on the waveform.

Patients with PLEDs may have myoclonic jerks that are either synchronous with the PLEDs or independent. When the jerks are independent, the generator for the myoclonus is probably deep. Even when they are synchronous, the generator is probably subcortical. The cortical discharge reflects projections from the deep generator.

Herpes Simplex Encephalitis

Herpes simplex encephalitis usually shows PLEDs on EEG during some phase of the illness, although at other times, there is slowing in the theta and subsequently delta range. The PLEDs are sharply contoured slow waves with a frequency of 2–4 Hz. The duration of each wave is often more than 50 ms. This relatively slow frequency of repetition helps to differentiate PLEDs in herpes encephalitis from the higher frequency discharges of subacute sclerosing panencephalitis (SSPE).

Neonates with herpes encephalitis may have necrosis that is not confined or even most prominent in the temporal region. These patients often do not have PLEDs. The EEG may show a poorly organized background with slow activity in the delta range predominating.

Anoxic Encephalopathy

Anoxic encephalopathy is also sometimes called *hypoxic–ischemic encephalopathy*, since the most common cause is cardiac arrest, where there is not only loss of oxygenation but also loss of blood flow to the brain.

The background is disorganized with diffuse slowing and suppression. Periodic sharp waves are often seen and may predominate in the record. They look similar to PLEDs, except that they are synchronous between the hemispheres. Patients may have myoclonus associated with the discharges. These probably represent the extreme of the burst-suppression pattern, seen often in patients with anoxic encephalopathy.

Burst-Suppression Pattern

The burst-suppression pattern occurs in patients with severe encephalopathies. The finding is not specific as to etiology but is most often seen in patients with hypoxic–ischemic damage and in barbiturate coma. The pattern has the appearance of epochs of relative flattening of the background with interspersed riefer epochs of mixed high-amplitude bursts (Figure 5.10). Some of the bursts have a very sharp appearance, although they do not have the stereotypic appearance of epileptiform activity.

The burst-suppression pattern in different clinical conditions can look very similar. In fact, the burst suppression pattern can look similar to the markedly discontinuous pattern of 29-week conceptional age newborns. Bursts of slow waves with superimposed sharp activity are superimposed on a very suppressed background. The background is not flat, but rather is very low voltage, composed of a mixture of frequencies.

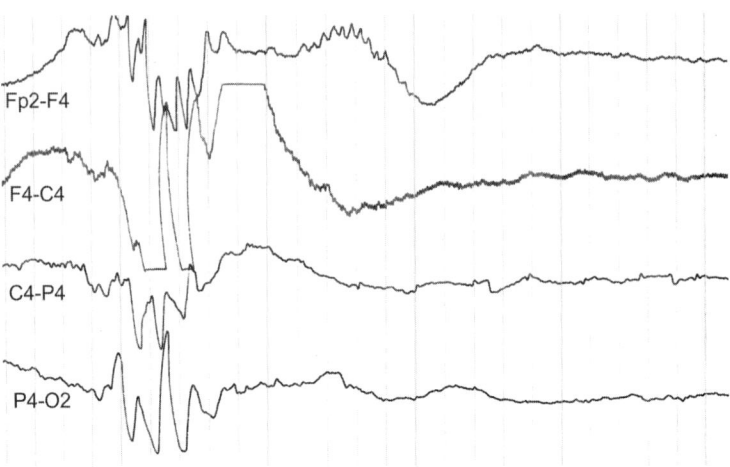

FIGURE 5.10 *Burst–suppression pattern, which some people call the suppression–burst pattern since the bursts occupy a lesser temporal proportion of the recording than the suppressions. Right medial portion of the LB montage.*

Subacute Sclerosing Panencephalitis

SSPE has almost disappeared as a result of measles immunization. Periodic complexes are seen in most patients at an intermediate stage. Early on, there may be only mild slowing, with disorganization of the background. Late in the course, the periodic complexes may completely disappear, leaving the recording virtually isoelectric. The discharges are slow waves with sharp components. The duration of the complex is up to 3 seconds, and the interval between complexes is 5–15 seconds. The background during the interval is disorganized and generally suppressed. Myoclonus is typically synchronous with the discharge.

EEG in SSPE resembles the burst-suppression pattern. The background is usually more suppressed with burst suppression than with SSPE. The two patterns are more easily differentiated by clinical presentation. Patients with burst suppression usually have a known history of hypoxia or severe metabolic derangement. Patients with SSPE have a typical history of a progressive neurologic disorder with intellectual deterioration and seizures. SSPE is very rare.

Creutzfeldt–Jakob Disease

Creutzfeldt–Jakob disease (CJD) produces an evolving pattern of EEG findings that depend on the stage of the disease. At some point in the disease process, a periodic pattern is seen, composed of a sharp wave or sharply contoured slow wave. The interval between discharges is 500–2,000 ms. The discharges are maximal in the anterior regions and may occasionally be unilateral (Figure 5.11). Only late in the disease are the discharges prominent posteriorly and when so are commonly associated with blindness. The discharges may or may not be temporally locked to myoclonus.

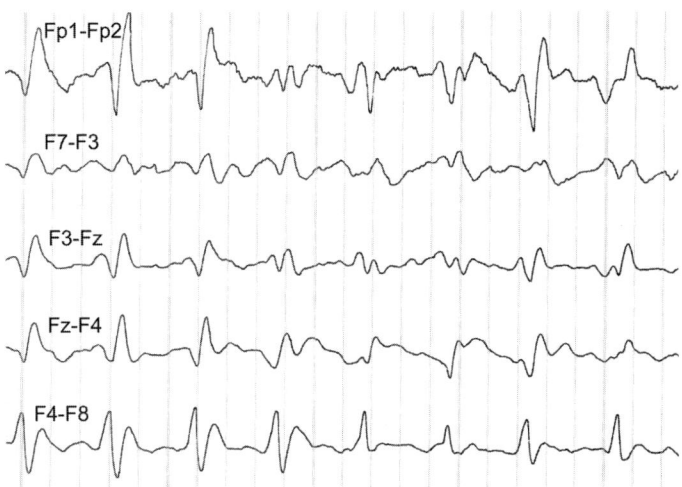

FIGURE 5.11 *Periodic discharges in a patient with Creutzfeld–Jakob disease. The discharges are seen from both sides. These are the frontopolar and frontal channels of the TB montage.*

These discharges are superimposed on an abnormal background characterized by low-voltage slowing in the theta and delta range. The periodic complexes abate in sleep.

Early in the course, the periodic complexes cannot be seen and the only finding may be focal or generalized slowing. About 10–15% of patients may not show periodic patterns during their course.

6   Neonatal Electroencephalography

Technical Requirements

Recording Procedures

Neonatal EEG is performed similarly to adult EEG but the montages are different and there is more physiological monitoring. The filter and sensitivity settings are the same as for adult EEG. Sensitivity commonly has to be reduced because of the high-voltage activity.

The *Guidelines* (American Electroencephalographic Society, 1994) recommend that at least three physiologic parameters be monitored:

- Respirations
- Eye movements
- ECG

Respiration can be rapid, producing an artifact that mimics slow activity on the EEG. ECG monitoring helps with identification of cardiac and pulse artifact. Eye movement recordings and submental EMG help to identify wake and sleep states.

Newborns spend most of their time sleeping, so sedation is usually not necessary. The electrodes are placed before the baby is fed. Then the baby usually quickly falls asleep after feeding. The baby should be aroused later in the study to observe arousal and the waking state.

Photic stimulation is of little benefit in newborns and is not routinely performed. Driving responses are not consistent and photoconvulsive discharges are rare at this age. Hyperventilation cannot be voluntarily performed and is not artificially performed.

Montages

Recommended montages for neonatal EEG are shown in Table 6.1. The most commonly used montage is the *Newborn montage*. This is a version of the longitudinal bipolar (LB) montage with fewer scalp leads; there are two channels between the frontopolar and occipital regions rather than four for

TABLE 6.1 Montages for neonatal electroencephalography

Channel	Longitudinal bipolar	Ear reference	Newborn montage
1	Fp1-F3	Fp1-A1	Fp1-C3
2	F3-C3	Fp2-A2	C3-O1
3	C3-P3	F3-A1	Fp1-T3
4	Fp2-F4	F4-A2	T3-O1
5	F4-C4	C3-A1	Fp2-C4
6	C4-P4	C4-A2	C4-O2
7	F7-T3	P3-A1	Fp2-T4
8	T3-T5	P4-A2	T4-O2
9	T5-O1	O1-A1	T3-C3
10	F8-T4	O2-A2	C3-Cz
11	T4-T6	T3-A1	Cz-C4
12	T6-O2	T4-A2	C4-T4
13	ECG/EMG	ECG/EMG	ECG/EMG
14	Resp	Resp	Resp
15	Left EOM	Left EOM	Left EOM
16	Right EOM	Right EOM	Right EOM

Resp = respiration; EOM = extraocular movement.

the adult LB montage. However, since the newborn head is smaller, the interelectrode distances are comparable.

The full adult LB montage can be applied, although the potential for smear between electrodes is high. Also, this should only be done on a 21-channel machine because of the physiologic monitoring that should be done. For a 16-channel machine, a reduced version of the LB montage can be used, as shown in Table 6.1. If there are extra available channels, recording from the vertex is desirable, for instance Fz-Cz, Cz-Pz, since this is not a routine part of many LB montages.

The montage is commonly changed during an adult recording, but we prefer to select a single montage for the entirety of a neonatal recording. This allows for easy detection of state changes. We routinely use the *newborn montage*.

Guidelines for Interpretation

Neonatal EEG is very different from adult EEG, so excellence in interpretation of the latter does not indicate competence in the former. Normal background, abnormal background, and epileptiform activity looks very different in the neonate. For accurate interpretation, the neurophysiologist needs to know:

- Gestational age
- Postnatal age
- Physiologic state
- Reactivity
- Clinical question

Conceptional age is used for interpreting neonatal EEG, Conceptional age is the sum of gestational age plus postnatal age. Maturation of the EEG has defined stages, which are presented below.

Physiologic state refers to the wake–sleep cycle. The cycles and terminology are different in adults and children. This is also discussed below.

Interpretation of the neonatal EEG requires a systematic approach:

- Examine the frequency composition and distribution of the background. Is the background appropriate for the conceptional age and physiologic state?
- Look for left–right asymmetries in the background. Is one side suppressed in comparison with the other? Does one side have excessive delta activity in comparison to the other?
- Look for sharp waves and spikes. Are they unifocal or multifocal? Unilateral or bihemispheric? Are they in the frontal or temporal region? Are they single or repetitive?
- Look for possible seizure activity. Are there episodes of suppression due to desynchronization? Is there a stereotypic rhythm? Monomorphic alpha in neonates is usually a subcortically generated seizure discharge.
- Look for changes in background with changes in state. An invariant pattern may be abnormal.
- Can a possible abnormality be explained by a normal rhythm?

Normal Neonatal Electroencephalography

Wake–Sleep Cycle

Normal term patterns of EEG activity are seen by 38 weeks conceptional age. At term, two stages of sleep are identified, quiet sleep (QS) and active sleep (AS). QS is characterized clinically by the absence of movement and regular respirations. AS is characterized by small eye and body movements and less regular respiration. AS is the equivalent of rapid eye movement (REM) sleep and QS is considered to be the equivalent of non-REM sleep.

Two EEG patterns are associated with QS. One is slow-wave sleep, in which continuous delta activity predominates, and the other is *tracé alternant* (TA), in which there are alternating periods of relative quiescence and bursts of sharply contoured theta activity. These bursts may be 3–6 seconds long. The interburst activity ranges from 5 seconds to almost 15 seconds. The TA pattern must be distinguished from a pathologic burst suppression and from the normal discontinuous pattern of premature infants. The EEG in AS is characterized by theta activity with some delta and beta activity superimposed. During the course of a long sleep, the first AS period is higher amplitude than subsequent AS periods. The later AS periods have more theta and less delta.

TABLE 6.2 Neonatal electroencephalographic maturation

Conceptional age	Electroencephalographic findings
22–29 weeks	Long periods of low-voltage activity with short bursts of higher voltage mixed-frequency bursts that contain sharply contoured theta with faster frequencies Interburst interval may be 2 min
29–31 weeks	Still a discontinuous pattern but interburst intervals are shorter Appearance of delta brushes
32–34 weeks	Discontinuous pattern in quiet and active sleep Appearance of multifocal sharp transients
34–37 weeks	Discontinuous pattern in quiet sleep but progressively shorter interburst intervals Active sleep (REM) is almost continuous Less multifocal sharp transients Appearance of frontal sharp transients
38–40 weeks	*Tracé alternant* pattern in non-REM sleep, with burst-to-interburst ratio of 1:1 May be a continuous slow-wave pattern Fewer frontal sharp transients

REM = rapid eye movement.

Maturation of the Electroencephalogram

The preceding description of the newborn EEG is only true for term infants. The EEG background matures quickly from 29 weeks to 38 weeks. In general, the background becomes more continuous, and wake–sleep states become distinct as the brain matures. Table 6.2 summarizes neonatal EEG maturation.

Conceptional Age 22 to 29 Weeks

The EEG shows long periods of low-voltage activity punctuated by short bursts of higher-voltage activity (Figure 6.1). The bursts are composed of mixed frequencies. Sharply contoured theta and faster frequencies can give the normal bursts an epileptiform appearance, but this is a normal pattern. The interburst intervals may last up to 2 minutes, although intervals of less than 1 minute are more typical. When the bursts first develop, there is poor synchrony between the hemispheres. With full development, there is good interhemispheric synchrony.

The alternating bursts and low-voltage activity are termed discontinuous; the pattern is called *tracé discontinu* (TD). This pattern appears similar to the burst-suppression pattern seen in some older patients with encephalopathy. The two are differentiated easily by knowledge of the conceptional age, but otherwise only with difficulty.

Conceptional Age 29 to 31 Weeks

The TD pattern is evolving with the interburst intervals shorter in duration and less regular. The interburst periods are of higher amplitude than in younger premature infants. Sleep stages are more differentiated than in younger neonates, and TD is seen prominently in QS.

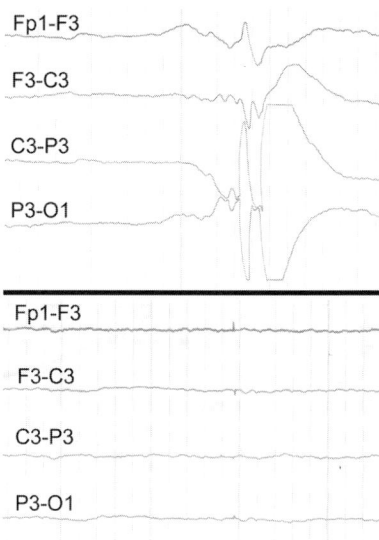

FIGURE 6.1 *Very premature infants have a markedly discontinuous EEG pattern. (Top) An epoch of EEG with a burst. (Bottom) An epoch of EEG without a burst.*

Delta brushes are prominent. These are composed of a delta wave with superimposed fast rhythmic activity in the alpha or beta range. Delta brushes are most prominent in the central and occipital regions and are best seen in AS. Delta brushes resemble sleep spindles but are physiologically different. Sleep spindles are not prominent in REM sleep and are minimal in the occipital regions. Also, the disappearance of delta brush later in development and the subsequent development of frontal spindles argue against a common physiologic substrate.

Conceptional Age 32 to 34 Weeks
The EEG during QS is still discontinuous, although the intervals of quiescence are shorter and less pronounced. Delta brushes are still present, and the spindle component is of higher frequency. Slow waves in the delta range are seen in posterior leads. AS is still discontinuous. Chin EMG is reduced during AS but is not a reliable indicator of state.

Multifocal sharp transients appear at this stage, occurring in the wake and sleep states. They are differentiated from pathologic spikes by their widespread distribution and lack of repetitive discharge.

Conceptional Age 34 to 37 Weeks
The EEG in QS (non-REM sleep) is still discontinuous but the interburst intervals are progressively shorter with increasing age. The burst–interburst time ratios are 1:2 and 1:3. AS (REM sleep) is virtually continuous, with delta predominating posteriorly and theta and faster frequencies anteriorly. For the first time, EMG becomes a reliable indicator of state, with low amplitude in REM sleep.

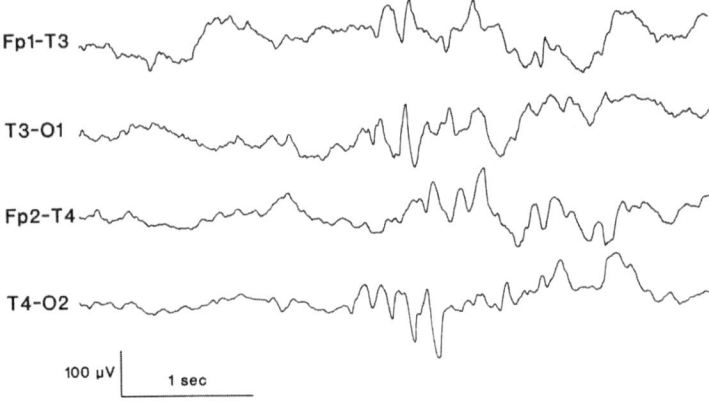

FIGURE 6.2 *Newborns at term often have a discontinuous pattern termed tracé alternant. The discontinuity is less prominent than in premature infants.*

Multifocal sharp transients are less prominent and are replaced by frontal sharp transients. These are of higher voltage than multifocal sharp transients. The EEG is more reactive to external stimuli than at earlier ages. Stimulation causes attenuation of the background and frequently is followed by a change in state.

Conceptional Age 38 to 40 Weeks
Term infants have good differentiation between REM sleep, non-REM sleep, and wakefulness. During non-REM sleep, the discontinuous pattern now has a burst–interburst ratio of about 1:1. This is the mature TA pattern (Figure 6.2). Non-REM sleep may be characterized by a continuous slow-wave pattern rather than TA. This pattern is occasionally misinterpreted as encephalopathy in neonatal EEG.

Frontal sharp transients are less prominent but may be seen until 2 months of age. Delta brushes are absent.

Abnormal Patterns

Abnormal neonatal EEG patterns fall into at least one of the following types:

- Abnormality of maturation
- Epileptiform activity
- Background abnormality

Lombroso developed a numeric classification of abnormalities, but this is not in widespread use. The classification scheme is presented on the CD.

Abnormalities of Maturation
Dysmature means that the EEG pattern is not appropriate for the conceptional age. For example, a discontinuous pattern with an interburst

interval of 1 minute is normal in a premature infant of 29 weeks conceptional age. This same pattern would be very abnormal in a term infant and would be indicative of encephalopathy. Persistent dysmaturity is associated with poor neurologic outcome. Transient dysmaturity may be due to non-neurologic causes and is not necessarily associated with brain damage.

Visual analysis of neonatal EEG allows for detection of only great discrepancies, but this is usually sufficient for routine interpretation. Quantitative analysis is possible but seldom needed and not in routine use.

Abnormalities of state are difficult to diagnose in routine EEG. An invariant pattern is abnormal, but state change may not necessarily be captured during a routine 20-minute EEG.

Background Abnormalities

Background abnormalities include:

- Excessive slow activity
- Low-voltage background
- Burst-suppression pattern
- Asymmetric patterns

Excessive slow activity is difficult to discern, since neonates have prominent delta activity already. Some infants with brain damage may have widespread delta, however. The slow background is present in wake and sleep states and reacts poorly to exogenous stimuli. This pattern is differentiated from normal delta activity by its widespread distribution and lack of reactivity. Normal delta is prominent anteriorally and attenuated by exogenous stimuli.

Amplitude asymmetries are significant only if they approach 50% or more. The asymmetry usually indicates focal cerebral damage in the region of suppressed voltage. A common pitfall is misinterpretation of asymmetries due to extracranial hematomas or fluid collections. Subdural hematomas may suppress activity from one or both sides.

The isoelectric EEG is a confirmatory test for brain death, in the appropriate clinical situation. Guidelines for determination of brain death are presented in Chapter 7.

The low-voltage record is unusual in neonates and suggests abnormalities in generation of electrical activity in the cortex. The technician needs to ensure that non-REM sleep is recorded, since normal REM sleep has a low-voltage background. Bilateral subdural hematomas may also produce bilateral attenuation of the background.

Epileptiform Activity

Epileptiform activity in the neonate can look very different from epileptiform activity in older children and adults. The epileptiform activity may be focal, multifocal, or generalized. Immaturity of cerebral maturation usually does not allow for generalization of epileptiform activity.

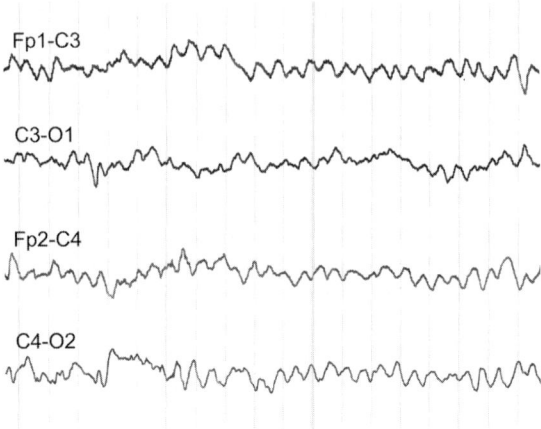

FIGURE 6.3 Rhythmic discharges in a neonate. Rhythmic activity in the alpha range is not normal during any stage of the sleep–wake cycle in a neonate. This is a seizure discharge. The generalized spike–wave patterns of seizures in older children and adults are not seen in neonates.

Focal discharges occur usually in the central region, more often on the right than the left. The discharges may occur singly or in trains at 5–10/sec (Figure 6.3). Focal epileptiform activity is differentiated from normal frontal sharp transients and multifocal sharp transients by consistent lateralization. Also, normal sharp transients never occur in trains. The focal discharges occasionally have a smooth contour and could be confused with an alpha or theta rhythm. Sustained rhythmic activity is never normal in neonates of any conceptional age, however. The rhythm must be differentiated from the fast component of delta brushes by the absence of an underlying slow wave and the longer duration of the epileptiform discharge than the fast component of a delta brush.

Focal discharges are usually associated with focal clonic seizures. The location of the focus may not necessarily correlate well with the clinical seizure activity. The prognosis for favorable neurologic outcome is good, since focal discharges in neonates do not necessarily indicate a focal structural lesion.

Most focal sharp waves are surface negative. Surface-positive waves are seen in some neonates with intracerebral hemorrhage. If the sharp wave is followed by a slow wave, the hemorrhage is most likely subarachnoid. If the sharp wave is not followed by a slow component, the hemorrhage may still be subarachnoid, but is more likely to be intraventricular, subependymal, or intraparenchymal. The specificity of positive sharp waves for hemorrhage is controversial, however.

Multifocal discharges are usually associated with an abnormal background, characterized by disorganization or suppression. The spikes can be single or multiple, occurring in trains similar to those of unifocal discharges. The prognosis for good neurologic outcome is poorer for multifocal discharges than unifocal discharges. Seizures are usually clonic and may be subtle. The chief differential diagnosis for multifocal discharges is normal multifocal sharp transients. The abnormal background is key to differentiation between these patterns.

Pseudo-beta-alpha-theta-delta is a descriptive term for a discharge that begins at 8–12/sec and gradually slows to 0.5–3.0/sec. The discharge may have a sharp appearance but alternatively may have a smooth contour. This is an ictal pattern, with typical seizures being tonic, myoclonic, or subtle. The pseudo-beta-alpha-theta-delta rhythm usually indicates a poor prognosis and is commonly seen in patients with perinatal asphyxia. The evolution of changing frequency is common, especially to frequencies that are a subharmonic of the original frequencies.

Rarely, neonates may manifest seizures without any perceptible alteration in background. The generator of epileptiform activity is probably subcortical, and the discharges are not projected to the surface. These infants usually have severe brain damage, explaining the lack of rostral projection of the activity.

7 Special Studies in Electroencephalography

Brain Death

Guidelines for Determination of Brain Death

The guidelines for determination of brain death (BD) are based on the consensus of the Medical Consultants on the Diagnosis of Death to the President's Commission for the Study of Ethical Problems in Medicine and Biomedical and Behavioral Research, hereafter referred to as the *President's Commission*. For BD, the patient must meet the following criteria:

- Cessation of all brain function
- Recovery not possible
- Known cause of coma

Clinical examination for BD should show the following:

- No pupillary reflexes
- No corneal reflexes
- No response to auditory or visual stimuli
- No response to "Doll's head maneuver"
- No response to ice water caloric testing
- No respiratory effort with apnea testing

The clinician must ensure that the absence of responsiveness is not due to drug intoxication, metabolic disturbance, or neuromuscular blockade. Therefore, the following parameters should be assured:

- Temperature at least 90°F
- Systolic blood pressure of at least 80 mm Hg
- No toxic levels of central nervous system (CNS) depressants
- No neuromuscular blockade

Patients being evaluated for BD will frequently be hypothermic and hypotensive; therefore, maintenance using warming blankets and pressors is often required. Evidence against neuromuscular blockade can be the presence of tendon reflexes, the presence of primitive responses to nociceptive stimulation, or the response of the muscle to electrical stimulation of motor nerves.

The guidelines for determination of BD indicate that there should be a period of observation, with documentation of examinations for BD before and after this period. If the cause of coma is not anoxia, the period of observation is 24 hours. Table 7.1 summarizes the brain death criteria.

The period of observation can be shortened if there is a confirmatory test. These tests include the following:

- EEG
- Brainstem auditory-evoked potential
- Radionucleotide blood flow study
- Angiogram

Recently, transcranial doppler (TCD) has also been studied as a confirmatory test for BD. Since BD is a complex legal issue and the President's Commission did not specifically mention TCD, this technique should not be used until the clinician can be assured that its use for the determination of BD is part of accepted medical practice.

TABLE 7.1 Brain death criteria

Criterion	Required data
Basic criteria for brain death	Cessation of all brain functions Recovery is not possible Cause of coma is known
No brain response upon clinical examination	Pupillary reflexes Corneal reflex Auditory and visual stimulation Oculovestibular reflexes ("Doll's eyes") Ice water caloric testing Apnea testing
Physiologic requirements for declaration of brain death	Temperature ≥ 90°F Systolic blood pressure ≥80 mm Hg No toxic blood levels of CNS depressants No neuromuscular blockade
Confirmatory tests	EEG BAEP Radionucleotide flow study Cerebral angiogram
Period of observation: Where cause is not anoxia	With confirmatory test: 6 hours Without confirmatory test: 12 hours
Where cause is anoxia	With confirmatory test: 12 hours Without confirmatory test: 24 hours

If a confirmatory test is performed and is consistent with BD, the period of observation can be reduced from 12 to 6 hours if the cause is not anoxia. The period of observation can be reduced from 24 to 12 hours if the cause is anoxia.

BD should usually be established by clinical findings alone, if possible. Some patients with no clinical evidence of cerebral or brainstem activity will have evidence of EEG activity but otherwise fulfill the clinical criteria for BD. The literature is not clear on what to do in this situation. The probability of meaningful neurologic recovery is virtually nonexistent if the patient has no evidence of cerebral or brainstem function throughout an appropriate period of observation, regardless of the results of a confirmatory test.

Guidelines for Brain Death in Children

The 1981 President's Commission did not make specific recommendations for the determination of BD in children. The only specific comment recommended "caution in children under the age of five years." The *Task Force for Brain Death in Children* (1987) subsequently provided recommendations that are increasingly used. The recommendations follow those outlines for adults except for the following:

- Do not declare a patient who is under the age of 7 days brain dead. Clinical and EEG criteria are not established for this early period.
- If the patient is between 7 days and 2 months of age, perform two examinations and two EEGs 48 hours apart to determine BD.
- If the patient is between 2 months and 1 year of age, perform two examinations and two EEGs 24 hours apart.
- If the patient is older than 1 year of age, perform two examinations 12 hours apart without a confirmatory test. This observation period can be 6 hours if a single EEG is done.

Despite these recommendations, most pediatricians do not feel comfortable with the diagnosis of BD without a confirmatory test. Table 7.2 shows a recommended montage for determination of brain death in children. Other parameters are similar to those used for determination of brain death in adults.

TABLE 7.2 Montage for determination of brain death

Channel	Montage
1	Fp1-C3
2	C3-O1
3	Fp2-C4
4	C4-O2
5	Fp1-T3
6	T3-O1
7	Fp2-T4
8	T4-O2

Electroencephalography for Brain Death

Technical standards include the following recommendations:

- Use a minimum of eight scalp electrodes covering all brain regions. This is usually a reduced version of the 10–20 Electrode Placement System. The following electrodes are recommended as a minimum: Fp1, Fp2, C3, C4, O1, O2, T3, T4.
- Use interelectrode distances of at least 10 cm. This allows for better detection of low-amplitude EEG activity. A minimal montage would be that shown in Table 7.2.
- Use interelectrode impedances that are no greater than 10 kohms but no less than 100 ohms. Too low an impedance occurs with electrode smear. The amplitude of recorded electrocerebral activity will be excessively low if the impedance is low.
- Use a sensitivity of 2 µV/mm during most of the recording.
- Use a low-frequency filter setting of 1 Hz and a high-frequency filter setting of 30 Hz.
- Use ECG monitoring and other physiologic monitoring if necessary. Monitoring of chest-wall motion may be needed if there is apparent slow activity in the record that might be respiratory.
- Record reactivity of the EEG to auditory, visual, and tactile stimuli.
- Use a recording time of at least 30 minutes, most of which must be relatively artifact-free recording.
- Test the integrity of the system by touching the electrodes to evoke a high-amplitude artifact. This ensures that a flat background is not due to technical factors.
- Telephone transmission EEG cannot be used to support the diagnosis of BD.
- Recording should be made by a qualified technician.

Brain Death Studies in Adults
BD studies should be performed in the period of observation between two extensive neurologic examinations. All of the above recommendations should be followed. All physiologic parameters set forth by the President's Commission should be followed regarding temperature, blood pressure, and absence of sedative and neuromuscular blockers.

Brain Death Studies in Children
BD studies in children are performed in the same manner as BD studies in adults. More physiologic monitoring is often required in children's studies, however. Because of small body size, respiratory movement artifact is relatively greater, and a chest-wall sensor is desirable. An ECG channel is desirable for adult studies but is even more important for BD studies in children. At high sensitivities, ECG artifact can be the predominant potential in the record.

Electroencephalographic Monitoring

Routine EEG is not always abnormal in patients with epilepsy. Therefore, long-term monitoring is often needed to help with the diagnosis. Just as long-term cardiac monitoring is helpful for evaluation of possible arrhythmia, long-term EEG monitoring can be invaluable. If a patient has episodes that are possibly seizures, but routine EEG is normal, EEG monitoring should be considered. Most fellowship-trained neurophysiologists have adequate training and experience, but physicians who are not trained should not attempt this interpretation. There are many potential pitfalls with these techniques.

There are two basic techniques: inpatient EEG monitoring and ambulatory monitoring. Although the latter is fine, and we have done this for years, we have largely abandoned this technique in favor of inpatient EEG monitoring. There is no substitute for being able to see the event on videotape or streaming media as we see the EEG. Nevertheless, both techniques offer invaluable information when making the important distinction between seizure and pseudoseizure.

Inpatient Electroencephalographic Monitoring

Methods

Monitoring Laboratory   There are three options for inpatient EEG monitoring:

- Dedicated EEG monitoring unit
- EEG monitoring capability in a sleep lab
- Portable EEG monitoring equipment placed in the patient's room

Dedicated EEG monitoring units are mainly the province of comprehensive epilepsy centers, where patients are monitored for days in preparation for surgery. These units are expensive but highly effective.

Most large hospitals have a sleep lab, and for a relatively small price, can be outfitted for EEG monitoring. In some cases, this only entails adding software to the existing sleep-lab equipment.

Portable EEG monitoring units are very helpful for patients in an intensive care unit, where there is concern as to whether electrical discharges are controlled when the patient may not manifest signs of obvious epileptiform activity. This equipment usually does not allow for recording of the appearance of the patient, but since the patient is observed by the nursing staff, a description of the event is typically reliable.

Electrodes and Montages   Electrodes and montages are the same as those used for routine EEG. In most circumstances, a complete set of electrodes is placed. Digital acquisition systems allow recording of each channel so that the montage and gain can be selected and changed during the reading by the physician.

Collodion is used rather than paste for these long-term recordings. If the recording exceeds 24 hours, the electrodes must be checked and gel reapplied to keep electrode impedances sufficiently low.

Performance

Many patients with pseudoseizures have multiple episodes per day, so that multiple days of recording are not necessary. We usually perform a daytime 6-hour recording if routine EEG is negative. If events are captured and can be interpreted, then further study is not needed. If not, then 24-hour recordings are needed. We often take patients off their anticonvulsants for these long-term recordings, which makes both epileptic seizures and pseudoseizures more likely.

Interpretation

Approach to Interpretation   Routine EEGs are completely reviewed by the physician; every page of every record. With the short duration of the recordings, this is easily feasible. However, long-term recordings cannot be completely examined, so there are time and event monitors on the recording system and the physician will review the EEG around each of the events. Also, spike detection software produces some regions of interest that the physician will evaluate.

The interpreting physician will typically review all of the regions of interest. Seizure discharges are reviewed along with segments of EEG before and after the discharge. Clinical episodes will also be reviewed, using side-by-side comparison of the EEG and video recording. Also, regions of abnormalities identified by the technician will be reviewed by the physician.

Seizure Discharges   Seizure discharges may be subtle and easily overlooked on prolonged recordings where there are multiple regions of interest. More commonly, the discharges are obvious; generalized or focal spike–wave complexes can appear before and during clinical seizure activity.

Difficulty in interpretation may occur when muscle artifact obscures the recording during the ictal event. Reducing the gain can eliminate some of this obscuration. Changing some filter settings can help accomplish this, as well. Unfortunately, changing filter settings can also alter the frequency response of the system, altering spike as well as artifact configuration.

Pseudoseizures   Most of the studies performed in most EEG monitoring labs are for suspected pseudoseizures. Any experienced neurologist will admit that history is often unable to accurately differentiate epileptic seizures from pseudoseizures. Even direct observation can be misleading. This difficulty is amplified by the fact that many patients can have both epileptic seizures and pseudoseizures. Figure 7.1 shows the results of EEG monitoring in some patients with pseudoseizures.

Ambulatory Electroencephalographic Monitoring

Ambulatory EEG monitoring is performed in many laboratories, but lacks the ability to observe the patient during epochs of electrophysiologic activity.

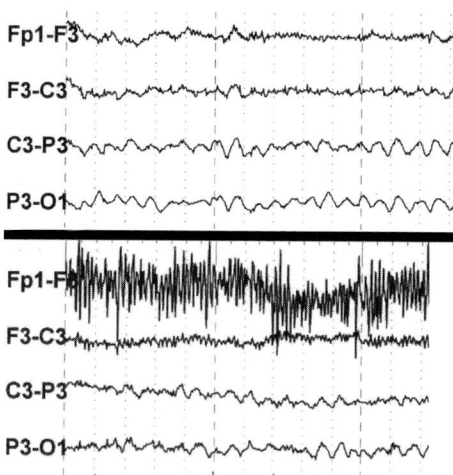

FIGURE 7.1   EEG activity of a patient with pseudoseizure. (Top) Normal awake pattern. (Bottom) Clinical seizure with muscle artifact from the frontal lobes and normal background from the central and posterior regions

The patient has an electrode array that is usually attached with collodion. The electrodes are connected to a recorder that is hung around the neck of the patient. Output from the electrode array is recorded usually on magnetic tape, although disks are sometimes used.

Quantitative Electroencephalography

Quantitative EEG has been a research tool for decades, but only recently has been available to the practicing clinician. However, the clinical applications are limited, so that QEEG is not part of routine practice. On first glance, digital signal acquisition lends itself to quantitative analysis, possibly making EEG interpretation more objective, and less dependent on the experience and bias of the interpreting physician. However, the calculations reduce the amount of data that is considered in interpretation of the information, thereby reducing some of the diagnostic power of the study. Also, the calculated findings may not be significant for the patient. As has been discussed elsewhere, amplitude asymmetries of up to 50% are normal for the waking posterior dominant rhythm. This magnitude of difference would stand out in digital frequency analysis yet be unimportant for the clinical interpretation of the record.

QEEG should be used as a tool that supplements rather than replaces conventional visual EEG analysis. There is no substitute for the analytical ability of nature's own neural networks; QEEG findings should be interpreted with perspective as to their mathematical origins and limitations.

Methods

Digital analysis begins with analog-to-digital conversion, as previously described. The information from each channel is converted independently then stored and manipulated. Montages are created by comparing the signal voltages from one channel with one or more others. For example, a longitudinal bipolar montage might be created by first subtracting the signal from F3 from the signal from Fp1, and so forth. Usually, the native channel data are stored along with a guide to the interpretive algorithms, so that these can be reselected at a later time. Filtering is performed with a broad frequency response.

Spike Detection

Spike detection is the most helpful feature of digital signal analysis. Spike detection software performs frequency and amplitude analysis on the EEG record to identify epochs that may contain epileptiform spikes. The software tends to err on the side of detection, so most highlighted events are not true spikes; artifacts and physiologic activity predominate. Nevertheless, the spike detection software allows the amount of EEG to be reviewed to be reduced considerably. The neurophysiologist should review the EEG before and after the spike to see if there is a state change, potential for artifacts, and potentially ictal events.

Power Spectral Analysis

Power spectral analysis was one of the first applications of digital EEG analysis. This involves separation of the EEG signal into fundamental frequencies, determining the amount of each frequency in the record. The data are displayed as power as a function of frequency.

Power spectral analysis is usually used in EEG for giving a visual impression of frequency content. This can be particularly helpful for determination of encephalopathy or sleep state. This can be used for intraoperative monitoring, especially. A relative increase in slow activity or suppression of all activity over one hemisphere during surgery is worrisome for infarction or some other insult to the underlying cortex.

Brain Mapping

Brain mapping is the display of frequency data topographically. Brain mapping is most helpful for detecting small asymmetries that would suggest a structural lesion. In this regard, digital analysis is more sensitive than visual analysis for detecting these subtle differences. Part of the limitation of brain mapping is that much of the data displayed are interpolated, calculated from few actual recording points. This gives a nice appearance, but may introduce data that did not exist. Minor differences may appear dramatic on the mapping yet be clinically unimportant.

Interpolation is automatically performed by the computer. Data from individual electrodes are used to calculate a predicted signal from points that were not recorded. For example, calculation of the potential at a point between F3 and C3 would give equal weight to the signal from these electrodes, but would also factor in contributions from Fz, Cz, F7, and T3, although at a lower weight.

Brain mapping is used mainly for patients with seizures and patients with dementia. In seizures, brain mapping can aid in the identification of areas of increased epileptiform activity indicative of a focus. In dementia, QEEG can increase the sensitivity of routine EEG to detect mild slowing suggestive of an organic dementia rather than pseudodementia. Cerebrovascular disease has been studied in QEEG because changes in EEG are immediate, whereas imaging abnormalities may not be evident for days. Although this is academically interesting, the EEG information does not currently have a clinical use in routine care of acute stroke or transient ischemic attack (TIA). Perhaps, in the future, EEG will help determine whether a patient requires thrombolytics, since the EEG may be different in persistently ischemic versus reperfused brain. However, magnetic resonance imaging is more likely to be used in the stroke centers.

Part III Nerve Conduction Study and Electromyography

8 Basic Principles of Nerve Conduction Study and Electromyography

Overview

Nerve conduction studies (NCS) and needle electromyography (EMG) constitute the routine neuromuscular evaluation. Routine NCS include:

- Motor NCS
- Sensory NCS
- F-wave study
- H-reflex

Other NCS are less commonly used, and include:

- Paired stimulation
- Repetitive stimulation at high and low rates
- Blink reflex

Needle EMG testing is performed in virtually all studies. The options for EMG are:

- Conventional needle EMG
- Macro EMG
- Surface EMG
- Single-fiber EMG

Other studies include the sympathetic skin response, and are included with special tests in a separate discussion.

NCS and EMG are used for a wide variety of indications, and an individual approach is usually needed. Some of the most common indications for study are:

- Focal or diffuse weakness
- Focal or diffuse numbness
- Muscle cramps

Some of the most common diagnoses reached after NCS and EMG are:

- Peripheral neuropathy
- Carpal tunnel syndrome
- Ulnar neuropathy
- Myopathy

Performance and interpretation of neuromuscular diagnostic studies requires a good knowledge of anatomy and physiology. The anatomy will be presented along with the presentation of techniques. Physiology is discussed below.

Neuromuscular Physiology

Normal Neuromuscular Function

We tend to think of the motor and sensory systems as separate entities, from motor and sensory peripheral axons to the motor and sensory cortical regions of the brain. Yet, the anatomical and physiologic separation of these systems is greatly reduced with ascension in the nervous system, and ascension in the evolutionary line. For this chapter, the motor and sensory systems are considered individually.

Motor Function

Descending input to the motoneurons in the spinal cord depolarizes the dendrites. This depolarization is conducted to the axon hillock where there is opening of voltage-dependent sodium channels. When there has been sufficient depolarization to establish an action potential, the efferent potential is conducted down the motoneuron axon to the neuromuscular junction. The action potential depolarizes the nerve terminal, which then causes release of neurotransmitter into the junction. Binding of acetylcholine to receptors on postjunctional muscle membrane produces depolarization of the muscle membrane. When the depolarization is sufficient to generate an action potential, the potential is conducted throughout the muscle fiber. This depolarization causes release of sequestered calcium, thereby facilitating muscle contraction. The contraction is terminated when calcium is taken up by the sarcoplasmic reticulum and recycled for another contraction.

Activation of one motoneuron results in activation of every muscle fiber in that motor unit. One motoneuron action potential results in one muscle fiber action potential for a wide range of firing frequencies. There is normally no spontaneous muscle fiber or motoneuron discharge, meaning without descending or reflexive activation.

Muscle fibers associated with muscle spindles are *intrafusal* and muscle fibers responsible for the power of contraction are *extrafusal*. These are innervated by separate motor axons. The muscle spindles provide feedback to ensure that the extrafusal fibers generate power for sufficient shortening. If the muscle does not shorten sufficiently with the contraction, the frequency of discharge of active units is increased and inactive units are recruited.

Motor axons are of different sizes, the axons for the fast-twitch (type II) motor units being larger than the slow-twitch (type I) muscle fibers. This difference in motor axon diameter is mirrored by a difference in diameter of the muscle fibers, with the fast-twitch fibers often being larger in diameter than the slow-twitch fibers. This size relationship is not exact, and differs between muscles, and between species. Physiologic recruitment is size dependent, with low-effort contractions generally recruiting smaller-diameter muscle fibers. Again, this is variable, and large-diameter muscle fibers can be recruited first in certain circumstances.

Sensory Function
Physiologic activation of sensory afferent nerves is accomplished in a variety of ways. There are free nerve endings and many types of specialized transducer organs. The common final path to sensory nerve activation is alteration of the conductance of the sensory nerve terminals, which then produces a generator potential in the nerve terminal. The depolarization of the peripheral sensory nerve then begins the action potential cascade, which culminates in generation of the afferent action potential. The action potential is conducted proximally to the spinal cord and brain.

Electrical stimulation of the sensory nerves produces activation of potentially all the afferent axons, regardless of sensory modality. However, the large-diameter afferents have the lowest threshold, whereas the small-myelinated and unmyelinated axons have the highest threshold to electrical stimulation. Therefore, the submaximal stimulation of evoked-potential studies typically does not activate these small fibers. Sensory NCS use maximal stimulation, so the entire spectrum of sensory afferents is stimulated. However, the fast-conducting large-diameter myelinated afferents contribute most to the sensory nerve action potential (SNAP), so the sensory NCS tends to measure the conduction of the fastest fibers.

Abnormal Neuromuscular Function

Neuromuscular function can be disturbed at various levels, including:

- Motoneuron or sensory neuron cell body
- Peripheral nerve axon
- Peripheral nerve myelin
- Neuromuscular junction
- Muscle

The physiologic effect of each of these is summarized below and should be considered when individual disorders are discussed.

Neuron Cell Body Dysfunction
Nerve cell body dysfunction is often due to neuronal degeneration, and the most important disorders are amyotrophic lateral sclerosis (ALS) and spinal muscular atrophy (SMA). We constantly lose neurons throughout our lives, but in these and other neuronal disorders, cellular death is accelerated. The neuronal membranes become leaky so that there is influx of ions that depolarize the neurons. This depolarization may occasionally reach

threshold, producing spontaneous action potentials in the neurons. In ALS, this is manifest as *fasciculations*.

The depolarization of the neuronal cell body results in activation of the voltage-dependent sodium channels, but these channels are also time dependent—they become inactive after a short open time. The channels cannot open again until the membrane potential is re-established, so if this does not happen, they cannot be activated again. Therefore, the motoneurons become electrically inactive, contributing to the weakness.

The influx of ions, especially calcium, into the neurons results in activation or enzymes including proteases and phospholipases that essentially digest the neurons from within. This causes neuronal death, so the denervated muscle fibers try to find innervation from surrounding surviving motor axons. Therefore, there are many more muscle fibers innervated by a single motor axon, which is manifest on EMG as *giant potentials*. The new nerve connections are not as fast or as consistent as the ones that developed in early life, so there is a variety of times from motor axon activation to muscle fiber activation, which is manifest on EMG as *polyphasic potentials*.

Denervated muscle fibers develop their own fluctuations in membrane potential that can occasionally reach threshold. The single muscle fiber action potentials are the basis for *fibrillation potentials* and *positive sharp waves*. The difference between these two patterns is in geometry of action potential generation and recording, so there is not a pathologic difference between these patterns.

NCS with motoneuron degenerations show reduced amplitude of the compound motor action potential (CMAP) with little change in nerve conduction velocity (NCV), since the conduction of the fastest fibers is little affected. Sensory NCS is normal.

Peripheral Nerve Axon Dysfunction

Axonal degeneration is the most common type of peripheral neuropathy, and there is a multitude of causes. The common feature is degeneration of the distal portion of the axon with denervation of the innervated muscle. This produces EMG changes that are similar to those described for neuronal degeneration, including polyphasic potentials, fibrillations, and positive sharp waves. While fasciculations can occur with axonal degenerations, this is not as common as with neuronal degenerations.

NCS with axonal denervation is characterized by reduced amplitude of the CMAP and SNAP, although the normal ranges of amplitude are so wide that amplitude, alone, may not be sufficient to detect an abnormality unless there is prominent axonal dropout. NCVs are often normal but may be slightly slowed due to secondary demyelination. Axonal damage results in partial unraveling of the myelin sheath. If an axon is still functioning yet sick, the sheath may be dysfunctional enough to slow conductions.

Peripheral Nerve Myelin Dysfunction

Demyelination of peripheral nerves has a narrower differential diagnosis than axonal degenerations. The most important demyelinating conditions are autoimmune, including acute inflammatory demyelinating polyradiculoneuropathy

(AIDP) (or Guillain–Barré syndrome, GBS) and chronic inflammatory demyelinating polyneuropathy (CIDP).

Damage to the myelin sheath of peripheral nerve produces prominent slowing of the propagation of action potentials. Normal axonal conduction is fast because the myelin sheath increases the impedance of the axonal membrane. Depolarization of the axon results in electrotonic conduction of the depolarization to the next node, or gap in the myelin sheath. The axon membrane at the node is capable of generating an action potential, but the membrane between the nodes, underneath the myelin sheath, is not. Electrotonic conduction is virtually instantaneous, compared with action potential propagation, so with myelinated axons, the action potential essentially skips from node to node. This conduction is much faster than action propagation down an axon. The myelin sheath greatly reduces the decay in electrotonic conduction (Figure 1.4) that would normally occur, facilitating the fast conduction.

Demyelinating conditions destroy the increased transmembrane impedance, which helps electrotonic conduction, so failure of conduction is common. This is seen on NCS as *conduction block*. The deterioration of the electrotonic conduction also causes marked slowing of conduction velocity. This slowing also results in an alteration in appearance of the compound action potential (CMAP and SNAP), termed *dispersion*.

EMG is often normal in patients with demyelinating neuropathy, especially early on. However, with persistent demyelination there is secondary axonal damage, resulting in all of the features of denervation, discussed above. The findings are less prominent, and formation of long-duration polyphasics and giant potentials is not expected.

Neuromuscular Junction Dysfunction

Neuromuscular transmission is very secure in the absence of pathology; there is one muscle fiber action potential for every axon action potential. However, with neuromuscular junction abnormalities, this synaptic security is compromised. The exact manifestation depends on the pathology. Reduced release of transmitter, as with botulism and Lambert–Eaton (myasthenic) syndrome results in failure of activation of the muscle fiber. This is manifest on NCS as reduced CMAP with normal NCV. Repetitive stimulation at a fast rate can produce facilitation of transmitter release, thereby increasing the response, an *incremental response*. Myasthenia gravis is due to reduced numbers of available acetylcholine receptors, so there is also failure of muscle fiber activation. Repetitive stimulation does not help conduction, since some of the available receptors may be unable to bind to another volley of acetylcholine. Therefore, repetitive stimulation results in successively less response. This is a *decremental response*.

Muscle Dysfunction

The most important myopathies are muscular dystrophies and inflammatory myopathies (polymyositis and dermatomyositis). Myopathies produce damage to the muscle fiber membrane. Changes in conductance to ions results in influx of sodium that can occasionally reach threshold. This produces fibrillation potentials and positive sharp waves that are indistinguishable from those of neuronal degenerations. However, the motor unit potentials differ. There is no reinnervation so there are no long-duration polyphasic potentials or giant potentials. Rather, each motor unit has fewer function muscle fiber potentials giving small motor unit action potentials. The conduction in the muscle fibers is dispersed, so that the muscle fiber potentials are somewhat separated. This gives rise to polyphasic potentials, but the duration is short, since the change in conduction is less than with axonal damage. Therefore, myopathic motor unit potentials are of small amplitude and short duration. They are called brief small-amplitude polyphasic potentials (BSAPPs).

Equipment

Equipment used in electrophysiologic assessment of nerve and muscle is similar in overall design and function. Most equipment is essentially a computer with interface card for signal acquisition and controlling a stimulator. The stimulator is a waveform generator controlled by a controller module in the computer. The acquisition equipment consists of amplifiers, wide-band filters, and is external to the computer itself. The output from the acquisition equipment is then fed to an analog-to-digital conversion module of the computer.

Nerve Conduction Basics

Nerve conduction studies in the upper limb typically include both sensory and motor. More proximal studies in sensory nerves are called mixed NCS when the nerve under study contains both sensory and motor fibers. The most common nerves tested in the upper limb are the median, ulnar, and radial nerves. In conducting the study, it is easier to do the motor study first. This allows rapid location of the most efficient stimulation sites on the nerves that when marked properly are then used for the subsequent sensory nerve studies.

There is a wealth of information to be gained from NCS. The most commonly sought information is derived from analysis of conduction speed and amplitude of the waveform itself. NCV increases more proximally in the nerve. Velocities are generally greater than 50 m/s in the upper extremity and greater than 40 m/s in the legs. Amplitudes of CMAP and SNAP vary with technique and should be checked against established normative values. Maximal NCV is a function of the largest nerve fibers that have the most myelin and longest internode length. Strictly speaking, some slowing of NCV can be seen with dropout of the larger faster conducting fibers. This may actually account for much of the slowing seen in mildly abnormal studies. Amplitude is a function of the total number of fibers stimulated and the synchronicity of the impulse. If the waveform is normal in appearance and low in amplitude, a dropout of axons is usually the culprit. This is seen in pathologic processes that primarily affect axons. If the waveform is much broader and has several phases, then demyelination is usually to blame, causing dispersion and conduction block.

Motor Nerve Conduction Study

Methods

Electrodes For surface stimulation, the stimulator electrodes are usually stainless steel and mounted 2–3 cm apart on a small hand-held two-pronged probe (Figure 8.1). By convention the cathode (negative pole) under which negative charges collect is black in color. The anode (positive pole) is denoted by a red color. In most situations the cathodes of the stimulating and recording electrodes are adjacent on the nerve being studied and the anodes face away from each other. Checking that the electrodes are set up "black-to-black" aids in troubleshooting if problems arise. Certain studies require needle electrodes for stimulation. In this case, the cathode consists of a small needle inserted into the skin near the nerve and the anode may be either a surface electrode or another needle. Much less stimulator current is required when needle electrodes are used because the impedance of subcutaneous tissue is lower than that of skin, and the stimulating electrodes are much closer to the nerve under study.

Electrode Position The stimulating electrodes are placed on the skin overlying the nerve at two or more sites along the course of the nerve. The recording electrodes are placed over the belly of the muscle, with the active electrode over the midbelly of the muscle, as close to the estimated end-plate site as possible.

Stimulus Characteristics Stimulation of the nerve being studied is accomplished using a brief burst of direct current. Stimulators are of two

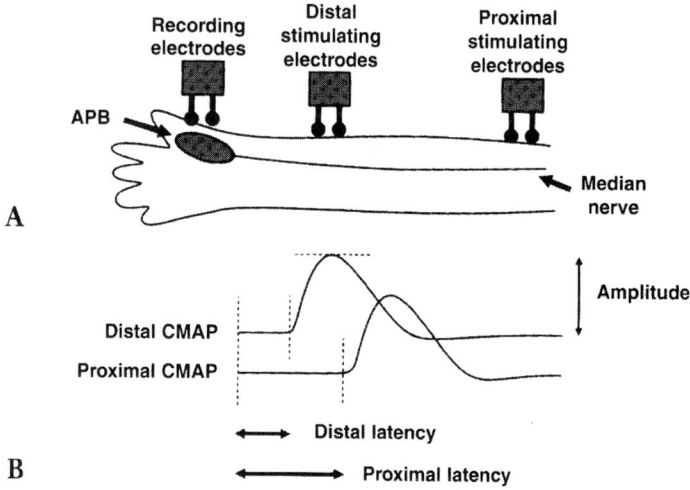

FIGURE 8.1 *Motor nerve conduction study.* **A.** *Diagram of the right forearm and representation of the electrode positions for median nerve conduction study. Photos of exact electrode positions are included on the companion CD.* **B.** *Sample recording of median motor compound motor action potentials (CMAPs).*

types: constant voltage and constant current. In our laboratory, constant-current stimulation seems to provide the most consistent responses. Constant-current stimulators vary the voltage of stimulation to compensate for changing skin impedance, providing a consistent current to the nerve being stimulated, while constant-voltage units vary the current to obtain consistent voltage to the nerve. Regardless of type, the amount of current applied to the tissue is never more than about 100 mA, and voltage rarely exceeds 500–600 V. Stimulus duration is usually in the range of 50–300 μs for most studies. One exception is the H-reflex study, which typically requires a much longer stimulus duration, on the order of 500–1000 μs, for best results.

Stimulus artifact can often be troublesome, especially in the case of sensory NCS. It can arise from numerous causes. Helpful in reducing stimulation artifact are: cleansing recording and stimulation skin sites with alcohol, placing the ground electrode between stimulation and recording sites, ascertaining that good electrical contact exists at all contact points, reducing stimulation intensity and/or duration, and increasing the distance between stimulation and recording sites when possible.

Procedure After all of the electrodes are in place, the instrument is set to deliver repetitive stimuli, usually at 1 Hz. The stimulus voltage is initially set to zero, then gradually increased with successive stimuli. A CMAP appears that grows larger with the increasing stimulus voltage. Eventually, further increases in voltage do not cause any change in CMAP amplitude. A stable response is assured if the voltage is 25% greater than the voltage needed to produce the highest amplitude CMAP. Once a good recording is made, the trace is stored for later analysis and the stimulating electrode moved proximally to a second stimulus site. Most nerves are stimulated in two sites for motor nerve conductions, but some are stimulated in at least three locations along the course of the nerve. It is not necessary to gradually increase the stimulus intensity for the subsequent electrode positions, since the patient is now used to the concept of electrical stimulation and would probably like to minimize the number of stimuli delivered. Table 8.1 shows common stimulus and recording parameters for NCS.

TABLE 8.1 Stimulus and recording parameters for motor and sensory nerve conduction studies

Parameter	Motor nerve conduction studies	Sensory nerve conduction studies	F wave	H reflex
Gain	2 mV/division	20 μV/division	200 μV/division	200 μV/division
Time base	2 ms/division	1 ms/division	10 ms/division	10 ms/division
Low-frequency filter	10 Hz	10 Hz	10 Hz	10 Hz
High-frequency filter	32 kHz	32 kHz	32 kHz	32 kHz
Stimulus duration	0.2 ms	0.1 ms	0.2 ms	0.2 ms

Interpretation

Measurements The following measurements are made from the CMAP produced by stimulation at each site:

- Latency from stimulation to onset of the CMAP
- Latency from stimulation to peak of the CMAP
- CMAP amplitude

If one of the waveforms is attenuated or there is a discrepancy in shape of the CMAP, this may have pathologic implications, but first ensure that this is not due to incomplete activation of the motor nerve.

The distance between stimulus sites is measured, and the NCV calculated according to the following simple formula:

$$NCV = \frac{Dist}{(PL - DL)}$$

where *Dist* is the distance, *PL* is the proximal latency, and *DL* is the distal latency. The final results are expressed as meters per second (m/sec).

Types of Abnormality The most common types of abnormalities in motor NCV are:

- Slow motor NCV
- Increased DL
- Relative slowing of motor NCV by comparing segments.
- Conduction block
- Decreased amplitude or altered waveform of the CMAP

There are defined norms for NCVs, and these are presented in Table 8.2.

Slow Motor NCV Slowing of the NCV below the normal range indicates a defect in the myelin sheath such that the axons do not conduct as fast as they normally should. This could be from neuropathy, nerve entrapment, cooling, and other causes. Comparing one NCS to others helps with this classification.

Increased Distal Latency Increased DL of the CMAP from the most distal site indicates slowing of conduction in the most distal portion of the motor nerve. This can be due to peripheral neuropathy, although in this case there would usually be slowing of conduction in more proximal segments of the peripheral nerve.

Relative Slowing of Motor NCV by Comparing Segments The absolute motor NCVs may be normal but there may be a discrepancy between the velocities of segments of the nerves. The best example of this is ulnar

TABLE 8.2 Normal data for motor and sensory nerve conduction studies

Nerve	Distal latency	Nerve conduction velocity	Amplitude
Motor nerve conduction studies			
Median	≤3.8 ms @ 7 cm	≥50 m/sec	≥5 mV
Ulnar below elbow	≤3.1 ms @ 7 cm	≥50 m/sec	≥5 mV
Ulnar across elbow	—	≥50 m/sec	≥5 mV
Radial	≤3.4 ms @ 6 cm	≥50 m/sec	≥5 mV
Peroneal	≤6.0 ms @ 8 cm	≥40 m/sec	≥2.5 mV
Tibial	≤5.0 ms @ 10 cm	≥40 m/sec	≥2.5 mV
Sensory nerve conduction studies			
Median	≤3.5 ms @ 13 cm	≥55 m/sec	≥10 µV
Ulnar	≤3.2 ms @ 11 cm	≥54 m/sec	≥10 µV
Radial	≤2.8 ms @ 10 cm	—	≥18 µV
Sural	≤4.2 ms @ 14 cm	≥42 m/sec	≥4 µV

entrapment across the elbow, where a difference in 10 m/sec in motor NCV is significant even with normal absolute NCVs.

Conduction Block   Conduction block is the extreme segmental slowing of motor NCV. This again indicates a defect in conduction of the nerves that can be related to nerve function or compression.

Decreased Amplitude or Altered Waveform of the CMAP   These abnormalities are not given the degree of importance that velocity changes are. Nevertheless, marked reduction in amplitude suggests axonal dropout. Severe damage to the myelin sheath can disperse the waveform so that the amplitude is less, but in general, amplitude correlates with axonal dysfunction more than myelin dysfunction.

Sensory Nerve Conduction Study

Methods

Electrodes   The electrodes are essentially the same as those used for motor NCV. In addition, spring-like stainless steel rings can be used for the fingers for stimuation or recording. The cathode is placed facing the stimulator. Certain near-nerve technique studies use a small needle inserted in the skin near the nerve as the recording surface. In sensory NCS the active and reference electrodes are placed along the nerve about 3–5 cm apart (Figure 8.2). Placing the ground between stimulus and recording sites can effectively reduce artifact. It is important to ask the patient to relax the limb being studied. This simple maneuver can especially increase the quality of sensory NCS.

Electrode Location   In order for the nerve response to be purely sensory, either the stimulating or the recording electrode has to be on a purely

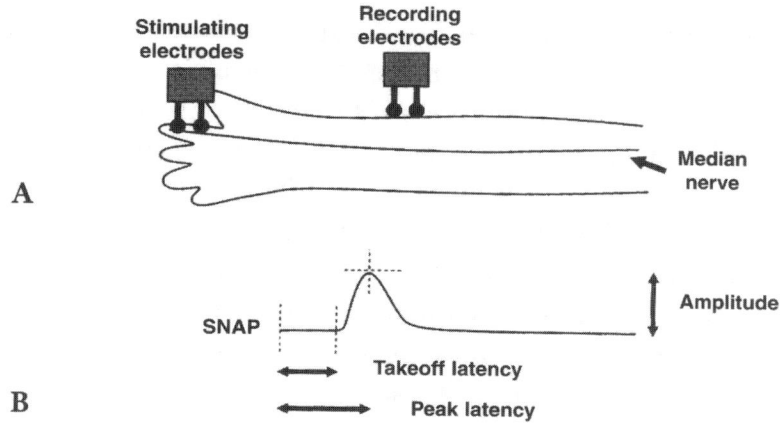

FIGURE 8.2 *Sensory nerve conduction study.* **A.** *Diagram of the right forearm for stimulation and recording of the median nerve sensory nerve action potential (SNAP).* **B.** *Sample recording of the SNAP.*

sensory portion of the nerve. For motor NCV, the recording is obtained from the muscle, so only motor fibers contribute to the response. With sensory studies, stimulation of a mixed nerve will activate both motor and sensory fibers so that recording has to be over a distal sensory branch of the nerve for a SNAP to be recorded. Alternatively, the stimulation could be reversed, so that the stimulation is on the distal sensory branch with recording from the proximal nerve trunk. Both ways are used. When the stimulus is distal, so that the nerves are conducting in the usual afferent direction, this is *antegrade conduction;* when the connections are reversed, this is *retrograde conduction*.

Averaging   Most modern equipment has the capacity to average the results of several stimulations. This is most helpful in sensory NCS where the signal-to-noise ratio is much lower and the typical working voltages are one or two orders of magnitude less than those seen in motor NCS. Use of the averager is quite useful in bringing out elusive low-amplitude sensory nerve action potentials.

Interpretation

Measurements   The following measurements are made:

- Latency to onset of the potential
- Latency to peak of the potential
- Amplitude of the potential
- Distance between stimulating cathode and active recording electrode

Sensory NCV is calculated from the following formula:

$$\text{NCV} = \frac{Dist}{LO}$$

where *Dist* is the distance between the stimulating and recording electrodes and *LO* is the latency to the onset of the sensory potential. Unlike motor NCS, both the onset and peak latencies can be used for interpretation, and most of us use peak latency at a defined distance as a measurement rather then calculating a conduction velocity.

Types of Abnormality The most common types of abnormality seen in sensory NCVs are:

- Slowed conduction of sensory nerves
- Low-amplitude response
- Absent response

Slowed Conduction Slowed conduction of sensory nerves, manifest as slowed sensory NCV or increased sensory distal latency, is due to damage to the myelin and can be focal as with entrapment or as part of a peripheral neuropathy. Comparison to motor conductions and other sensory nerve conductions will make this distinction.

Low Amplitude The amplitude of the response is normally quite low, such that averaging is often needed in order to get a reproducible and measurable response. Therefore, little significance is given to reduced amplitude as long as the latencies and conduction velocities are normal.

Absent Response Absent response is an extreme of reduced amplitude. However, absent response has to be considered abnormal. While one might assume that this means axonal damage, this is not certain, since modest myelin dysfunction can produce loss of the ability to record a reliable sensory response.

Nonpathologic Factors Affecting the Results

Age Velocity of NCS in term infants is generally half that seen in adults. Within 30–36 months of age, the NCS has reached normal adult speeds. Amplitude of NCS tend to be slightly higher in children due to lower skin impedances. After 60 years of age, NCS velocities begin to slow slightly.

Body Temperature Temperatures higher than 37°C do not produce any significant effect on NCS. Temperatures lower than 34°C induce two important changes: NCV slowing and increased amplitude. The latter is due to longer open time of the sodium channels when cooled. In the EMG laboratory, skin temperature should be measured prior to study and the limb warmed to a surface temperature of at least 32°C prior to beginning. Elaborate and expensive warming equipment is available to accomplish this task and is in use in many laboratories. We use a simple, efficient and cost-effective alternative: a stainless steel bucket filled with warm (40°C) water. We find that this warms the hand or foot faster and more completely than expensive infra-red heat lamps without any danger of burns or electrical

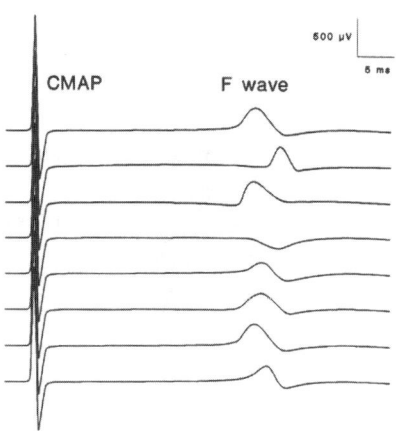

FIGURE 8.3 *F-wave study. Responses to motor stimulation with a slow time-base allow recording of the F-wave. The latency is taken as the minimum latency of the beginning of 10 consecutive F-waves. The F-wave is generated by retrograde activation of the motor neurons and then reflects off the motoneuron to conduct back down to the motor axon.*

mishaps. When warming to acceptable temperature is not possible, the NCV slowing can be corrected by adding 5% to the conduction velocity for each degree below 34°C There is no known correction for the augmentation in amplitude induced by cooling.

F-Wave Study

The F-wave tests conduction of motor axons proximal to the stimulation site. When motor nerves are stimulated for routine NCS, the stimulation creates action potentials that travel not only orthodromically toward the muscle but also antidromically toward the motoneuron. The antidromic potential reaches the soma and depolarizes the dendrites. Depolarization is then conducted electrotonically back to the axon hillock, which is now repolarized. A new action potential is created and transmitted back to the muscle. The action potential activates the motor end-plate, causing action potentials in muscle fibers. This late response is called the F-wave.

For the F-wave study, the electrodes are placed as for the motor NCS (Figure 8.3). The stimulating electrode is over the nerve either proximally or distally. However, the stimulating electrodes are turned around so that the cathode (negative pole) is toward the spine.

Normal values for F-wave conduction are presented in Table 8.2. The shortest of 10 successive F-waves is taken as the point to be measured. The latency is measured from stimulus to the onset of the F-wave, rather than the peak.

H-Reflex

The H-reflex is the electrophysiologic counterpart of the Achilles tendon reflex. The H-reflex is elicited by stimulation of the tibial nerve in the popliteal fossa while the patient is lying prone (Figure 8.4). The feet may be supported by a pillow to give a slight bend at the knee. Recording is made using surface electrodes over the soleus or medial gastrocnemius. The stimulus intensity is gradually increased and the H-reflex appears at a lower intensity of stimulation than the CMAP. As the stimulus intensity is further increased, the CMAP appears and the H-reflex disappears. The muscle afferents are larger and have a lower threshold for electrical activation than the alpha motoneurons, hence they are activated at a lower threshold of stimulation.

Normal H-reflex latency is 35 ms or less, and interside differences should not exceed 1.4 ms. H-reflex amplitude differs widely between patients and should not be used as a criterion, if the potential is visible and of normal latency.

Blink Reflex

The blink reflex can be used to evaluate patients with lesions of the facial nerve or of the brainstem. However, neuroimaging is the preferred method to evaluate the brainstem, so the blink reflex is used predominantly for evaluation of cranial nerves 5 and 7.

Methods Surface electrodes are placed as shown in Figure 8.5 (detail). Stimulus and recording parameters as well as interpretive information are shown in Table 8.3a and 8.3b.

Stimulation of the facial nerve produces a CMAP from the orbicularis oculi. This is termed the *direct response* and is a test of the integrity of the efferent system. Stimulation of the supraorbital nerve evokes the *blink reflex*. There are two components to the blink reflex: R1 and R2. R1 is a short-loop reflex that is only projected ipsilateral to the stimulus. R2 is a longer loop reflex projected bilaterally.

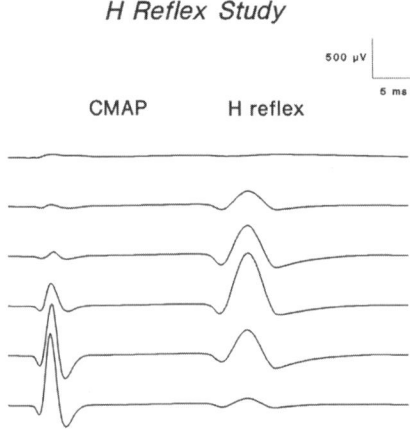

FIGURE 8.4 *H-reflex study elicited by stimulation of the tibial nerve and recording from the soleus. The stimulus intensity is gradually increased. The H-reflex is elicited first by activation of the muscle afferents. The afferents synapse in the spinal cord and response returns in the motoneurons. With further increase in stimulus intensity, the motoneurons are directly activated, which results in the CMAP and blocks the H-reflex.*

TABLE 8.3a Blink reflex

Parameter	Direct response	Blink reflex
Gain	1 mV/division	500 µV/division
Time base	1 ms/division	2 ms/division
Low-frequency filter	20 Hz	20 Hz
High-frequency filter	10 kHz	10 kHz
Stimulus duration	0.1 ms	0.1 ms

TABLE 8.3b Interpretation of blink reflex data

Direct	R1	R2 ipsilateral	R2 contralateral	Lesion location
Normal	Normal	Normal	Normal	Normal
Prolonged	Prolonged	Prolonged	Normal	Facial nerve
Normal	Prolonged	Normal	Normal	Brainstem pathways
Normal	Prolonged	Prolonged	Prolonged	Trigeminal nerve, but could be brainstem

Blink Reflex: Electrode Placement

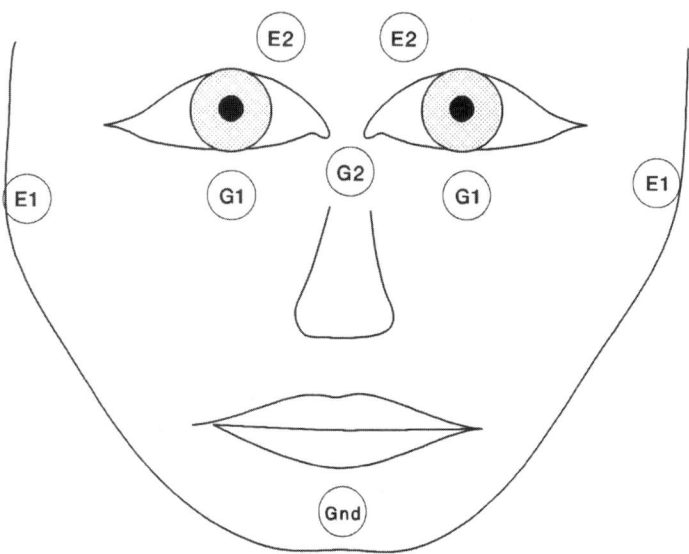

FIGURE 8.5 *Placement of electrodes for performance of the blink reflex. E1 is for direct stimulation of the facial nerve. E2 is for stimulation of the supraorbital branch of the trigeminal nerve. G1 is for recording from the orbicularis oculi. G2 is reference for both recording electrodes. Gnd is ground.*

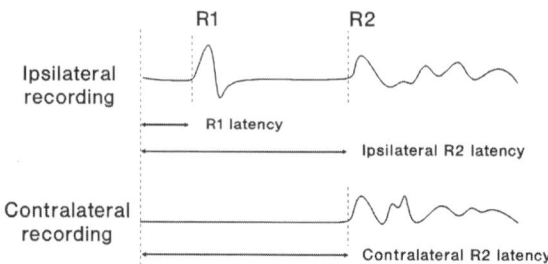

FIGURE 8.6 *Normal blink reflex. The ipsilateral recording shows a normal R1 and R2 response. Contralateral recording lack the ipsilateral R1 component. Latency is measured; amplitude is not of interpretive value.*

Measurements are made of the latency of the direct response, R1, ipsilateral R2, and contralateral R2. Amplitudes are too variable to be of interpretive value (Figure 8.6).

Interpretation   Interpretation of the blink reflex is as follows:

- Prolonged direct response with otherwise normal latencies indicates a lesion of the facial nerve, such as Bell's palsy.
- Prolonged R1 indicates a lesion in the reflex pathway from trigeminal nerve to facial nerve. Facial nerve lesions also prolong R1, but are distinguished from brainstem or trigeminal nerve lesions because the direct response is prolonged, as well. If the contralateral R2 is normal, then the afferent limb in the trigeminal nerve is normal, and a brainstem lesion is likely.
- If the latency of the direct response is normal but the latencies of R1, ipsilateral R2, and contralateral R2 are prolonged, a lesion of the trigeminal nerve is likely, but brainstem lesions cannot be excluded.

Electromyography Basics

Methods

Electrodes   In needle EMG the recording surface is typically the tip of a small sharp needle inserted through the skin into the muscle under study. In the case of monopolar EMG needles, the reference and ground electrodes are small surface electrodes of the type used in NCS. In the case of concentric needles, the reference electrode is the barrel of the needle and the active electrode is part of the needle tip electrically distinct from the barrel. Monopolar electrodes are used predominantly, nowadays.

Electrode Position The reference and ground surface electrodes are placed on the limb being studied, so that there is minimal opportunity for leak current to flow from one electrode through the heart and spinal cord to the others. When another limb is studied, the surface electrodes are moved also. The needle electrode is inserted into the belly of the muscle under study, near the expected region of the motor end-plate. Depending on the muscle, this is either midway through the length or at the junction between the proximal and middle thirds of the muscle. In general, the end-plate region is the thickest part of the muscle belly. Figure 8.7 shows electrode placement for some of the most commonly studied muscles. Table 8.4 shows common stimulus and recording parameters for performance of EMG.

Procedures The surface electrodes are placed first, then the needle inserted into the muscle to be studied. Then the amplifier is turned on to begin the recording. Recordings should be made in the following categories:

- Rest
- Insertion
- Single motor unit activation
- Maximal contraction

Figure 8.8 shows some patterns of normal EMG activity. Table 8.5 describes normal EMG activity.

Rest Normally, there will be little electrical activity in muscle. The electrode is moved, which will evoke a brief discharge, described below ("Insertional Activity"), then the muscle will be silent.

Insertion Since the amplifier is not turned on until the electrode is in the muscle, the initial insertion is not seen. However, as the electrode is moved to various sampling spots within the insertion tract, brief discharges will develop. These potentials are generated by individual muscle fibers that are mechanically stimulated by the insertion. Remember that the diameter of even a fine needle is vastly larger than that of a muscle fiber (imagine driving a car through a crowd of closely packed people); there will be copious electrical activity from those in the vicinity, even from cells not directly touched by the electrode. Deformation of the muscle fiber membrane produces local depolarization that then results in action potentials.

Single Motor Unit Activation After assessment of rest and insertion activity, the patient is asked to make a slight contraction of the studied muscle. This will evoke motor unit potentials (MUP) that are generated by each motor axon activating many muscle fibers. The motor unit may innervate 100–2000 muscle fibers, depending on the muscle tested.

Maximal Contraction Increasing force of contraction results in recruited units discharging at a faster rate and new units being recruited. This eventually produces a massive discharge with maximal contraction in which the

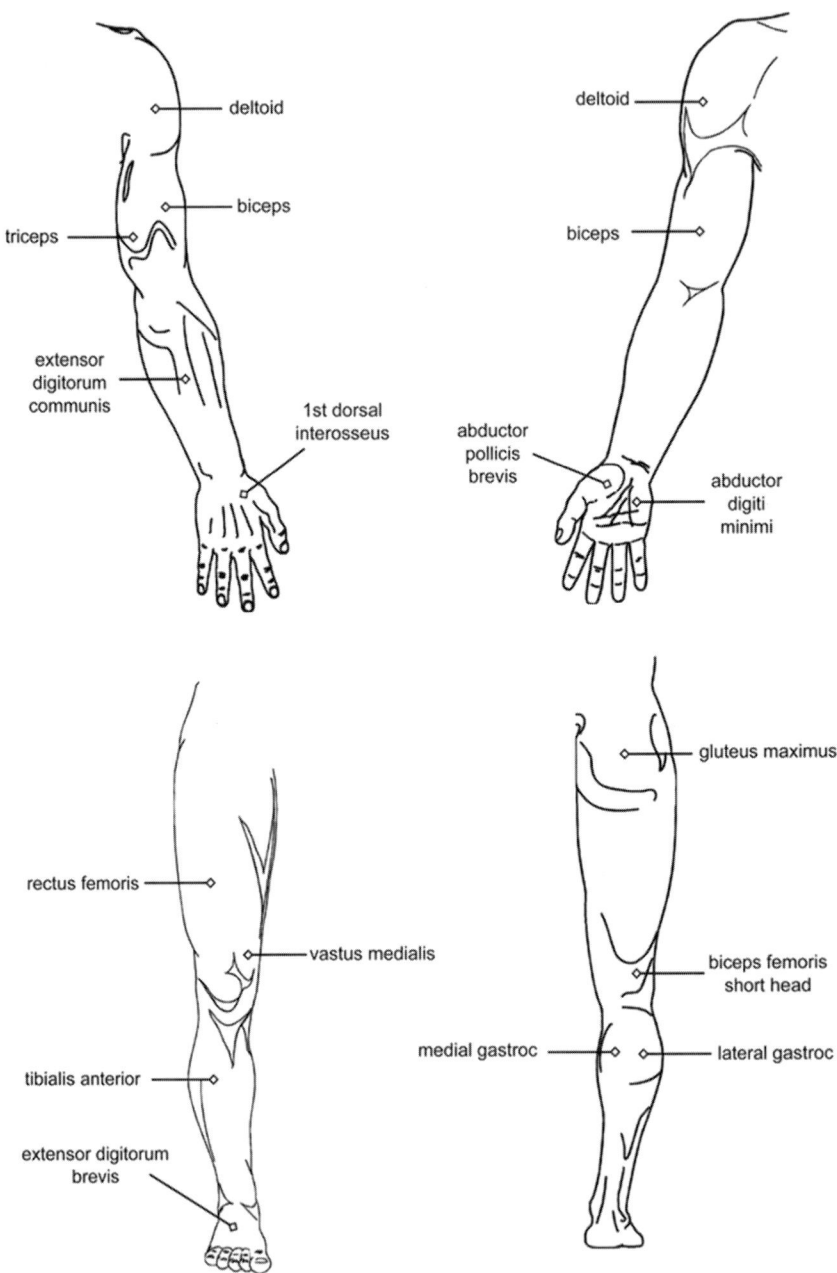

FIGURE 8.7   *EMG needle electrode positions for some commonly tested muscles. The electrodes are inserted perpendicularly to the skin.*

baseline is not visible and individual units are not discernible. This is the *interference pattern*. The term interference comes from the physical derivation rather than the football derivation—interference is the complexity of a waveform due to the interaction between two or more base frequencies.

TABLE 8.4 Stimulus and recording parameters for electromyography

Parameter	Resting	Motor unit	Recruitment	Single fiber
Gain	50 μV/div	200 μV/div	1 mV/div	0.2–1.0 mV/div
Time base	10 ms/div	10 ms/div	10 ms/div	0.1–1.0 ms/div
Low-frequency filter	10 Hz	10 Hz	10 Hz	500 Hz
High-frequency filter	32 kHz	32 kHz	32 kHz	32 kHz

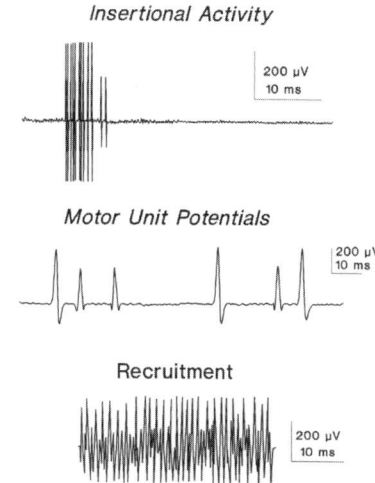

FIGURE 8.8 *Normal EMG patterns. Insertional activity is a brief burst of potentials. Resting activity is absent. Motor unit potentials (MUPs) are brief biphasic and triphasic potentials. Recruitment with vigorous contraction results in activation of many MUPs at fast rates, obscuring the background.*

TABLE 8.5 Normal electromyographic potentials

Pattern	Recorded	Findings
Resting activity	Muscle relaxed and needle not moving	No activity
Insertional activity	Movement of the needle in an otherwise relaxed muscle	Brief volley of action potentials
Motor unit potentials	Needle is not moved while patient makes slight contraction	A few motor unit action potentials, biphasic or triphasic, short duration
Recruitment	Patient makes progressively stronger muscle contraction until reaching maximum force	Increase number of functioning movements until the baseline is obscured

Muscles Chosen for Study The number and location of muscles chosen for study depends upon the clinical question. In the case of myopathy and proximal weakness, proximal muscles in at least two limbs are tested. The choice of muscles for EMG study should always be tailored to the specific clinical situation of each individual patient. Extreme care should be exercised when performing needle EMG in patients currently being treated with anticoagulant medication. In our laboratory we seldom perform EMG when the INR is greater than 2.0 or the PTT greater than 45.

Interpretation

Normal and Abnormal Responses

Figure 8.9 shows some commonly encountered abnormal EMG activity. Table 8.6 describes common abnormal EMG potentials. Table 8.7 shows the differential diagnosis of some common neuromuscular disorders based on neurophysiologic findings.

Resting Activity Normal resting muscle is electrically silent. There will be small amplitude of noise, but no sign of muscle fiber or motor unit activation. Many disorders of peripheral nerve and muscle produce instability of the

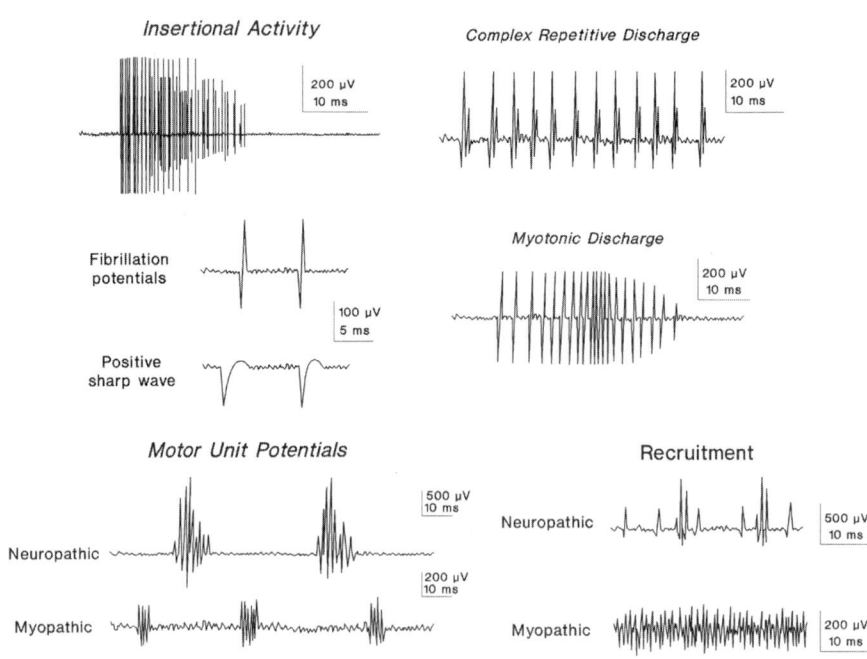

FIGURE 8.9 Abnormal EMG activity. Insertional activity is longer than in normal muscle, with a prolonged discharge especially in patients with myopathy and active denervation. Fibrillation potentials and positive sharp waves are spontaneous discharges of single muscle fibers at rest. Myotonic discharges have a distinctive sound on the audio monitor. Motor unit potentials can be neuropathic (long-duration, high-amplitude polyphasics), or myopathic (brief small-amplitude polyphasic potentials). Recruitment can be reduced, with neuropathic lesions, or early with myopathic lesions.

TABLE 8.6 Abnormal electromyographic potentials

Finding	Appearance	Physiology	Cause
Resting activity			
Fibrillation potentials	Brief biphasic potentials that occur at an irregular rate	Muscle fiber action potential due to membrane instability	Neuropathy, myopathy
Positive sharp waves	Positive sharp potentials with a fast downstroke and slower return to baseline	Muscle fiber action potential, as for fibrillation potentials	Neuropathy, myopathy
Fasciculation potential	Appearance of a motor unit, normal or neuropathic, discharging at an irregular rate, at rest	Spontaneous discharge of a motor unit	Neuropathy, motoneuron disease Normal in some patients
Complex repetitive discharge	Repetitive discharge of several muscle fibers that waxes and wanes	Repetitive discharge of muscle fibers	Chronic denervation Some patients with myopathy
Myokymia	Repetitive single motor unit potential discharges at 30–40/sec	Repetitive discharge of a single motor unit due to neuron membrane potential instability	Multiple sclerosis, brainstem glioma, radiation plexopathy
Myotonic discharge	Repetitive discharge of muscle fibers with a "dive bomber" sound in the audio monitor	Repetitive discharge of muscle fibers	Myotonic dystrophy, myotonia congenita, paramyotonia congenita, hyperkalemic periodic paralysis,
Motor unit activity			
Neuropathic motor unit potentials	Polyphasic potentials that are of long duration and usually high amplitude	Reinnervation causes activation of muscle fibers that discharge at disparate latencies	Neuropathy, neuronal degeneration
Myopathic motor unit potentials	Polyphasic potentials that are of short duration and small amplitude	Number of muscle fibers activated with each action potential is less and the time to fiber potential generation is variable	Myopathy
Maximal contraction			
Reduced recruitment	Rapid discharge of a few motor units that often have a polyphasic appearance	Fewer motor units activated with contraction; each unit discharges more rapidly	Denervation
Early recruitment	Many small motor units recruited at a low degree of tension	Decreased numbers of muscle fibers in each unit, and ineffectual contraction requiring activation of more small units	Myopathy

muscle fiber membrane. This results in episodic degeneration of action potentials that then is manifest as single muscle fiber action potentials. These are *fibrillation potentials* and *positive sharp waves*, discussed above. They have an amplitude of about 50 μV and are less than 3–5 ms in duration.

TABLE 8.7 Differential diagnosis of neuromuscular disorders

Diagnosis	Fibrillations	Fasciculations	Neuropathic motor unit potentials	Myopathic motor unit potentials
Acute denervation	Present, though not necessarily early	None	None	None
Chronic denervation	May be present	May be present, especially with neuronal degenerations	Present	None
Myopathy	Present	None	None	Present
Neuromuscular transmission defect	Absent	Absent	None	May be present

Occasional MUPs may be seen, and when they occur at a regular rate, are probably due to incomplete relaxation and, if otherwise normal in appearance, should not be interpreted as abnormal. However, irregular discharge of MUPs, with amplitude of more than 100 µV, suggests *fasciculations*. These are often polyphasic, suggesting denervation, but may not be.

Myokymia is an involuntary repetitive discharge of a single motor unit with a frequency of 30–40 Hz. A rippling appearance can be clinically observed. Myokymia usually results from irritation of the proximal nerve axon or ephaptic conduction in proximal portions of the nerve. It is usually observed in multiple sclerosis, brainstem glioma, and radiation-induced plexopathies. Gold-induced neuropathy is another cause.

Complex repetitive discharges (CRD) were once known as bizarre high-frequency discharges and this is a fitting description for the sound. The onset and offset is usually very abrupt and the sound is harsh and mechanical. This wave pattern is thought to arise from discharge of a specific muscle fiber that acts as a pacemaker for surrounding fibers. CRDs may occur spontaneously at rest or be evoked with insertion.

Insertional Activity Normal insertional activity is a brief burst of sound accompanying needle movement that sounds similar to cloth ripping. The sound should cease when needle movement ceases or very shortly thereafter. *Increased insertion* is when each needle movement produces a marked and prolonged discharge of the muscle fibers. *Decreased insertion* is when needle movement fails to evoke a prominent discharge of muscle fibers. This usually means that there are few functioning muscle fibers that are still electrically excitable. Complete absence of insertional activity is more likely due to a faulty electrode than a pathologic process.

Needle insertion may evoke *myotonic discharges*. These are repetitive muscle fiber action potentials. The pattern of repetition is characteristic and has a waxing and waning quality. This suggests to some a "dive bomber" quality on the audio monitor, but is more like a motorcycle accelerating and decelerating. The waveform itself is initially a negative slow onset followed by a faster positive component, similar to the positive sharp wave. A number of conditions can cause myotonic activity, discussed below.

Motor Unit Potentials   The physiology of normal and abnormal motor units was discussed earlier in this chapter. Normal motor units should be biphasic or triphasic with a duration of 10 ms or less. Longer duration units suggest motor unit reorganization or reinnervation. Polyphasic potentials also suggest reinnervation. If the polyphasic potentials are normal or of large amplitude and long duration (>10 ms) this suggests reinnervation. If the polyphasic potentials are of short duration and especially of low amplitude, this suggests myopathy.

Maximal Contraction   The recruitment pattern with maximal contraction is seldom the only sign of abnormality. *Decreased recruitment* means that there are fewer functioning motor units, so there is not the complete obscuration of the baseline that should occur. *Early recruitment* means that the baseline is obscured at low levels of effort. This is because there is ineffectual contraction of sick muscle fibers so that more motor units and muscle fibers are recruited.

Myopathy

In myopathy the muscle fibers are smaller and have less membrane area. Parts of the muscle fiber are incapable of supporting an action potential. Neuromuscular junctions (NMJs) may be unstable or dysfunctional because of inflammation or degeneration of the membrane at or near the NMJ. As a result of these processes, myopathic motor units are lower in amplitude, shorter in duration, and tend to be polyphasic and unstable. The characteristic sound produced on EMG is very scratchy. It sounds a little like sandpaper. Motor unit recruitment is abnormally rapid, with full interference pattern produced with relatively little effort.

Neuropathy

In the case of pure demyelinating neuropathy, no consistent abnormality is seen. In the case of axonal dropout and reinnervation, neuropathic potentials can be unstable. With continued chronic reinnervation, MUPs become high amplitude, long duration, and have a higher percentage of polyphasic potentials than normal. A special case is reinnervation after nerve injury. The newly forming nascent MUPs can be indistinguishable from myopathic units because of their small size and polyphasic waveform. Clinical history is essential to avoid misclassification of such findings.

Tests of Neuromuscular Junction Function

Repetitive Stimulation

In the normal NMJ, every depolarization of the nerve terminal leads to depolarization of the muscle membrane. There is a natural decrement in the total amount of acetylcholine released with each subsequent nerve depolarization down to a certain point. Even after long-term high-frequency firing of the nerve, the amount of neurotransmitter released is still far in excess of the minimum necessary for adequate NMJ signal transmission. This gap between what is available and the minimum necessary is called the safety factor. Disease states that interfere with either the release of neurotransmitter or its subsequent binding to the postsynaptic receptors decrease or eliminate the safety factor, leading to failure of neuromuscular transmission. Total failure of all NMJ transmission of course leads to total paralysis. Gradual failure of NMJs in the same muscle leads to weakness of that muscle. A single vesicle of acetylcholine in the presynaptic terminal contains about 50,000 molecules. These are randomly released and their resulting potentials can be recorded as miniature end-plate potentials (MEPPs). Diseases that interfere with neurotransmitter release decrease the frequency of MEPPs but not the amplitude, since amplitude of the MEPP is primarily dependent upon the number of available receptors on the postsynaptic membrane. Conditions that affect the postsynaptic receptors lead to a normal MEPP frequency but reduced amplitude. Repetitive nerve stimulation (RNS) testing reveals deficits in NMJ transmission by exposing the effects of an abnormal safety factor in the diseased state. Nerves are chosen for stimulation and CMAP amplitude is recorded at baseline and during a train of five or six stimuli delivered at 2–5 Hz. In some cases it is useful to record the result of stimulation at high rates of 50–60 Hz, requiring 50–100 stimulations. While the information gained from such testing is quite useful, the test is uncomfortable and is no longer offered in many laboratories. Similar information can be gained from examining pre-exercise and postexercise RNS tests. A decrement in amplitude between the first and last stimulation of greater than 10% is indicative of abnormal NMJ function.

Single-Fiber Electromyography

Single-fiber EMG (SFEMG) is a difficult and time-consuming technique performed primarily in academic centers. It generates two types of information: *jitter* and *fiber density*. Special equipment and recording needles are required. Filter settings are slightly different than those used in routine EMG. The needle is inserted into the muscle and low-level constant effort is supplied by the patient, sufficient to activate one motor unit near the recording needle. The trigger potential is the MUP nearest the needle, and the secondary potential is another recordable potential on the screen, which is part of the same motor unit. In a perfect NMJ, the difference between firing times of these two potentials will be very low. If one or the other NMJ is diseased in some way, the interpotential interval will vary, making the secondary potential seem to "jitter" back and forth on the screen. The extreme expression of abnormal jitter is blocking, in which case the secondary potential fails to appear after the primary potential fires, indicating a transient total breakdown of NMJ transmission of the secondary potential.

Fiber density is a measure of the total number of active fibers of the same MUP within a distance of 300 μm. The finding of increased fiber density is relatively nonspecific. Increased fiber density can be seen in myopathies because muscle fiber diameter is decreased, and remaining fibers are closer together. It can be seen in axonal neuropathies and radiculopathies because of reinnervation. It can be seen in advanced age because of decrease in overall fiber diameter, leading to packing of more fibers into a smaller space.

Evaluation of Individual Nerves

The approach to peripheral neuropathy should include sensory and motor NCS in at least one arm and one leg. Motor and sensory nerves in each limb should be tested, preferably in at least two segments, a more proximal one as well as a distal one. EMG should be done in at least one proximal and one distal muscle in each limb. This allows an opportunity to gauge the extent of the axonal involvement and also to look for conduction block and any degree of demyelination. In the western hemisphere the most common cause of neuropathy is diabetes mellitus. Diabetes can cause a wide range of abnormalities including axonal neuropathy, demyelinating neuropathy, and a mixed picture. Electromyography is sensitive and helpful in identifying acute and chronic denervation changes associated with axonal involvement in neuropathy. It should be considered an essential part of a complete electrophysiologic assessment.

Generalized neuropathies can be primarily axonal, primarily demyelinating, or a combination of both.

Mononeuropathy is extremely common in clinical practice. While almost any nerve can suffer entrapment or injury at one time or another, some are seen more commonly than others and we will confine our discussion to those most likely to be encountered in the typical electrophysiology laboratory.

Median Nerve

Median nerve anatomy is summarized in Figure 8.10. Median motor NCS is performed by placing the active recording electrode on the midbelly of the abductor pollicis brevis (APB) over the thumb. The reference is placed 2 cm distal. The cathode for distal stimulation is 7 cm proximal to the active recording electrode near the wrist crease. The proximal site is over the median nerve proximal to the antecubital fossa. In both instances, the anode is placed 2 cm proximal to the cathode.

Median sensory NCS is performed by recording from one of the fingers, usually the index, with ring electrodes while stimulating 13 cm proximally, above the wrist crease. The stimulating and recording electrodes can be reversed, and the results may differ slightly depending on the direction of conduction.

The most common entrapment neuropathy seen in our laboratory is carpal tunnel syndrome (CTS), which is a distal median neuropathy within the carpal tunnel at the wrist. Clinical diagnosis is usually straightforward, and the characteristic electrophysiologic findings no less so. The most common finding is slowing of distal sensory NCV and/or prolonged median nerve terminal latency, in the absence of abnormalities in other tested nerves. It is commonly bilateral and usually worse in the dominant hand. Denervation in the APB is sometimes seen in severe cases and warrants more aggressive

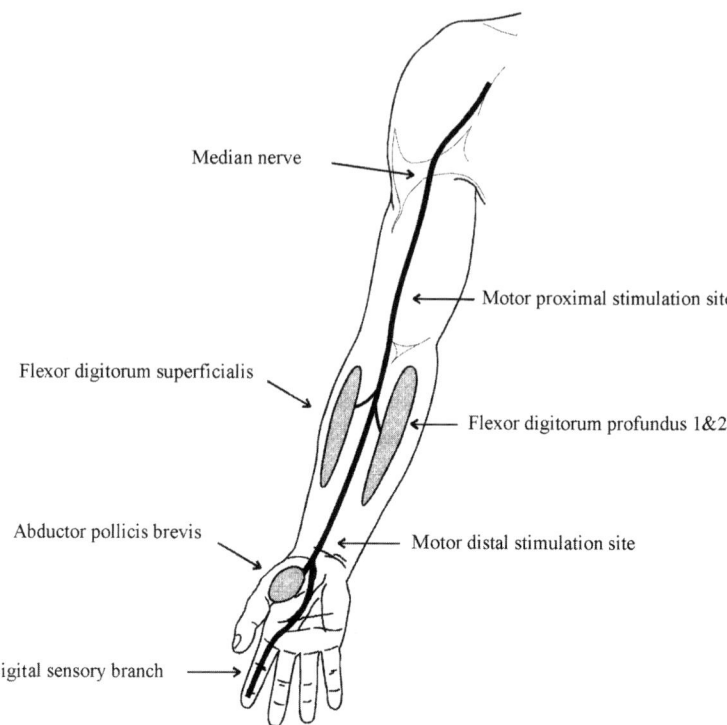

FIGURE 8.10 *Anatomy of the median nerve with important muscles shown.*

treatment, typically consisting of surgical decompression of the nerve in the carpal tunnel.

Ulnar Nerve

Ulnar nerve anatomy is summarized in Figure 8.11. Ulnar motor NCS is performed by placing the recording electrode on the midbelly of the abductor digiti minimi with the reference electrode 2 cm distal. Distal stimulation is delivered to the ulnar nerve 7 cm proximal to the active recording electrode, overlying the ulnar nerve in the distal forearm. Proximal stimulation is performed usually in two locations: below the ulnar groove at the elbow, and 10 cm proximal to that point. This provides two complete segments for assessment of nerve conduction.

Ulnar sensory NCS is performed by recording from digit 5 using ring electrodes and stimulating the ulnar nerve in the distal forearm 11 cm proximal to the active recording electrode.

Ulnar neuropathy at the elbow is another commonly seen entrapment syndrome. Slowing across the elbow in motor NCV and abnormalities in the sensory NCS are the usual findings. Occasionally, the lesion can be primarily axonal in nature and very little slowing is seen but the EMG shows denervation in the ulnar-innervated hand intrinsic muscles. Denervation is not seen in the flexor carpi ulnaris because the nerve branch innervating this muscle leaves the ulnar nerve proximal to the elbow.

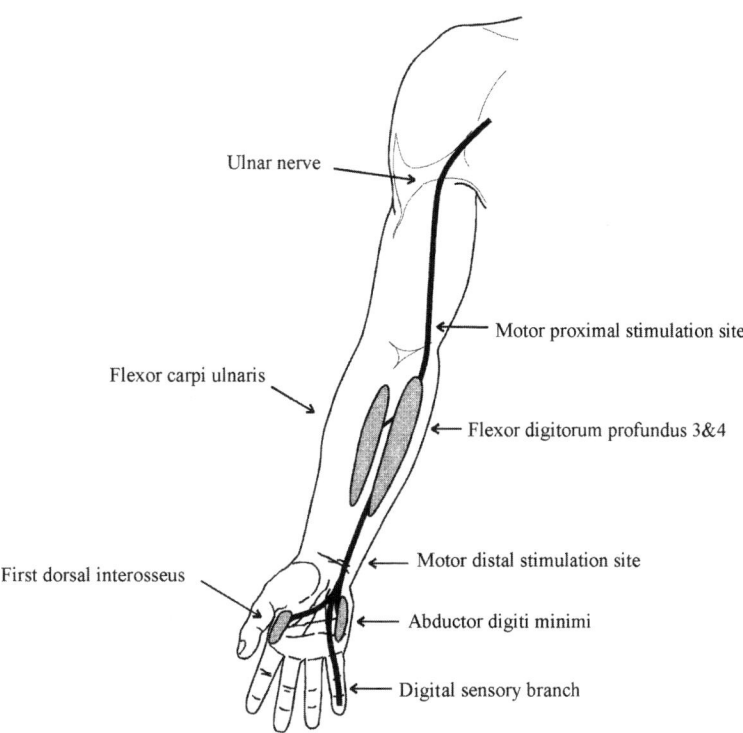

FIGURE 8.11 *Anatomy of the ulnar nerve with important muscles shown.*

Radial Nerve

Radial nerve anatomy is summarized in Figure 8.12. Motor NCS is performed by recording from the extensor indicis. The active recording electrode is placed over the belly of the muscle and the reference is placed 2 cm distal. The distal stimulation site is in the forearm, between the ECU and extensor digitorum minimi, 10 cm proximal to the styloid process. Proximal stimulation is just proximal to the antecubital fossa, between the biceps tendon and the brachioradialis. Sensory NCS is performed by recording from the superficial sensory branch on the dorsum of the hand while stimulating from one of the more proximal sites, listed just above.

Radial nerve entrapment or damage in the spiral groove in the humerus is not rare and typically follows a night of overindulgence. This is the so-called Saturday night palsy. Wrist drop and finger drop are seen along with loss of sensation in the distribution of the radial sensory nerve. Abnormalities in radial sensory NCS are seen and in motor NCS. The amount of denervation in radial muscles seen on EMG is variable. This entity is easily distinguished from C7/C8 radiculopathy by demonstrating normal function in median and ulnar-innervated muscles.

Peroneal Nerve

Peroneal nerve anatomy is summarized in Figure 8.13. Peroneal motor NCV is usually performed by recording from the extensor digitorum brevis. Distal

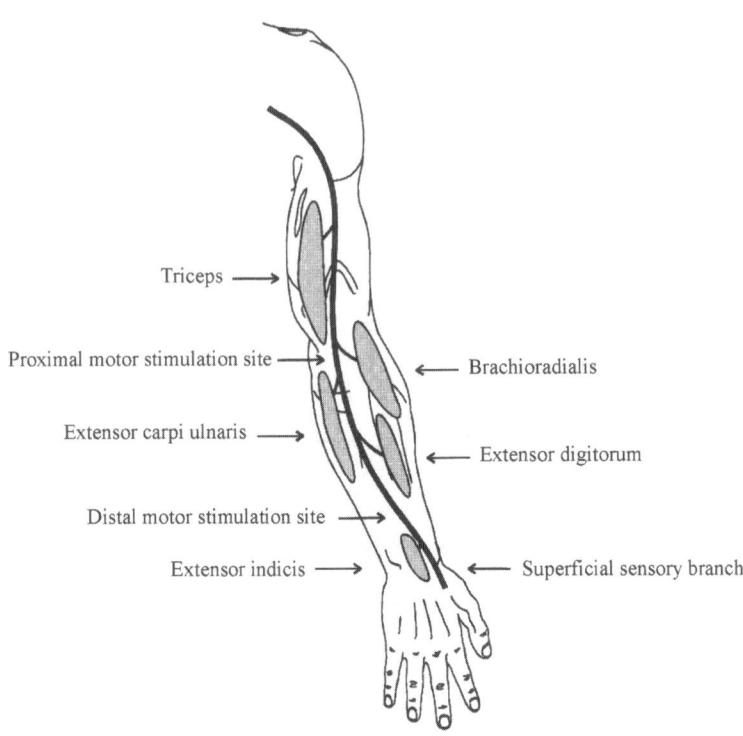

FIGURE 8.12   *Anatomy of the radial nerve with important muscles shown.*

stimulation is in the lower leg, adjacent to the tendon of the tibialis anterior. Proximal stimulation is at the fibular neck. When the nerve is thought to be injured across the fibular neck, proximal stimulation is then performed in the popliteal fossa. The NCV across the fibular neck is compared to the NCV distal to the neck. A difference of greater than 10 m/sec is abnormal.

Peroneal nerve entrapment at the fibular head producing foot drop is the most common entrapment syndrome of the lower extremity seen in our laboratory. Slowing of peroneal NCV at the fibular head is the classic finding. Denervation in tibialis anterior and peroneus muscles is variably seen. Care must be exercised to exclude the possibility of L5 radiculopathy and sciatic neuropathy. In the case of the former, denervation and weakness in posterior tibialis and hamstring muscles is seen in addition to findings in muscles supplied by the common peroneal nerve. In sciatic neuropathy, an abnormal EMG in the biceps femoris short head is sufficient to establish a more proximal localization of injury.

Tibial Nerve

Tibial nerve anatomy is summarized in Figure 8.14. The tibial nerve innervates the medial and lateral gastrocnemius and soleus muscles and supplies sensation to those portions of the sole and dorsolateral foot that are not

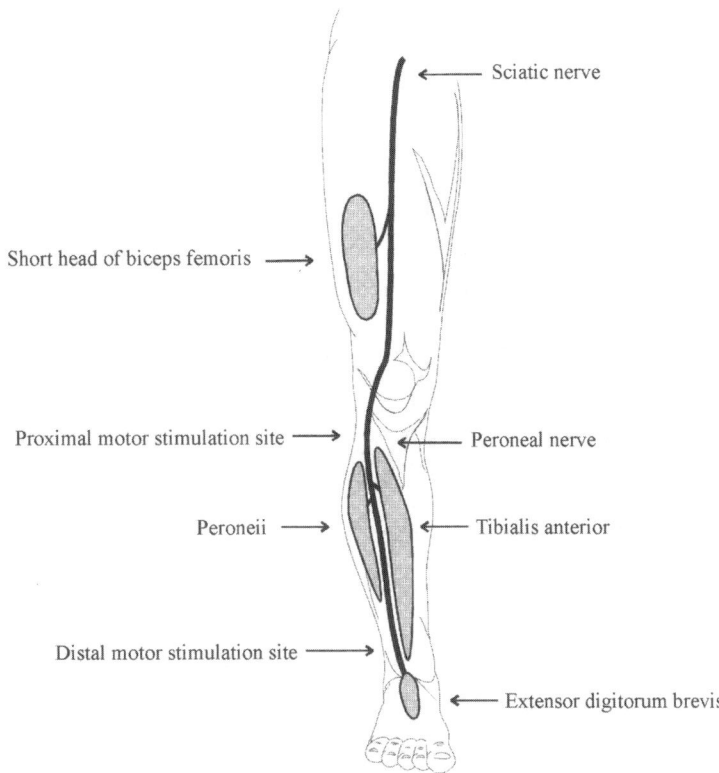

FIGURE 8.13 *Anatomy of the peroneal nerve with important muscles shown.*

served by sural and superficial peroneal nerves. The tibial nerve also innervates most of the intrinsic muscles of the foot via the medial and lateral plantar nerves. Tibial sensory NCVs are rarely performed and will not be discussed. Motor NCVs are performed by recording from the belly of the abductor hallicus muscle on the medial aspect of the foot. Distal stimulation is delivered to the nerve as it passes behind the medial epicondyle, and proximal stimulation is delivered in the popliteal fossa.

Sural Nerve

The sural nerve is the only purely sensory nerve in the leg that is tested routinely. The sural is formed in the midcalf by the joining of branches from both the peroneal and tibial nerves. Sural sensory conduction is recorded from the nerve behind and slightly inferior to the lateral epicondyle. The stimulation site is on the posterior surface of the leg, 14 cm proximal to the recording site. There are no definite landmarks for the site of proximal stimulation, so some hunting may be needed. For most other NCS, the ground is placed proximal to the stimulating electrode to minimize the potential for current to pass through the body; however, for sural NCS,

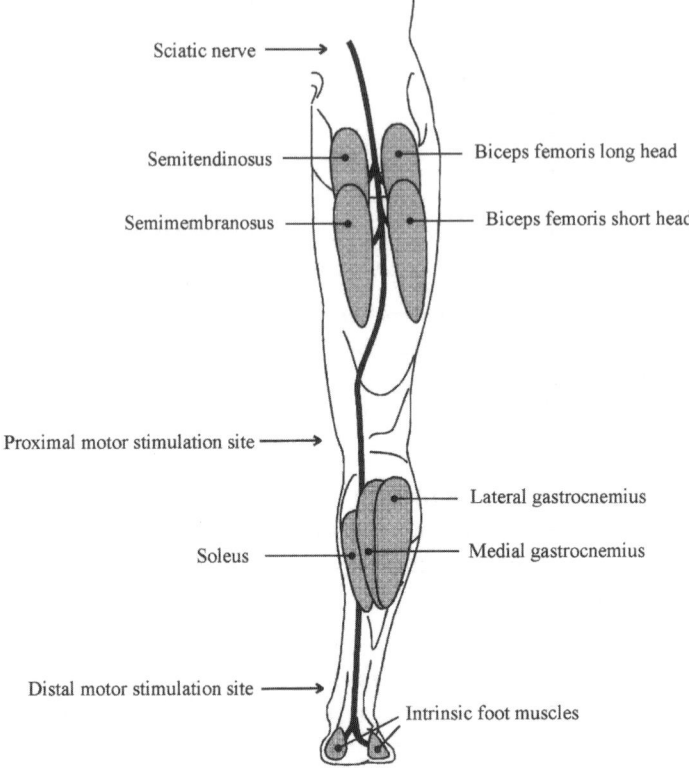

FIGURE 8.14 *Anatomy of the tibial nerve with some important muscles shown.*

the ground is often placed between the stimulating and recording electrodes to minimize noise.

Sciatic Nerve

Sciatic neuropathy is seen less commonly but often can masquerade as peroneal neuropathy. In most sciatic nerve injuries the peroneal division bears the brunt of the injury with relative sparing of the posterior tibial division. The reason for this is unclear. EMG revealing denervation in the distribution of both divisions of the sciatic nerve is helpful in establishing this diagnosis. NCS are not performed on the sciatic nerve, per se, but rather on the individual divisions: peroneal and tibial.

Femoral Nerve

The femoral nerve is supplied by roots from the lumbar plexus and L2-L4 make up most of the innervation. NCS of the femoral nerve can be technically challenging, so most of the diagnosis rests on EMG. Denervation in the quadriceps suggests a femoral neuropathy.

Evaluation of Individual Muscles

This section is presented in tabular form for conciseness, since neurophysiologists must have an extensive knowledge of the anatomy of individual nerves. Table 8.8 presents the innervation and action of important muscles that will be tested in EMG.

Evaluation of Neuromuscular Transmission

Three principal conditions constitute 99% of all cases of NMJ transmission defects seen in the average EMG laboratory: myasthenia gravis (MG), Lambert–Eaton myasthenic syndrome (LEMS), and botulism. MG is a postsynaptic defect, and LEMS and botulism are presynaptic defects with distinct pathophysiology. Defects in NMJ function cause weakness of affected muscles with no sensory loss. In addition to abnormal results in RNS testing, increased jitter in SFEMG is seen with all NMJ disorders. The types of abnormalities on RNS differ, depending on the disorder.

TABLE 8.8 Innervation and action of important muscles

Muscle	Nerve	Plexus	Root	Action
Arm				
Abductor pollicis brevis	Median	MC	C8, T1	Abduct thumb
Biceps	Musculocutaneous	LC, UT	C5-6	Flex elbow
Deltoid	Axillary	PC, UT	C5-6	Abduct arm
Extensor digitorum communis	Radial	PC, MT, LT	C7-8	Extend middle finger
First dorsal interosseus	Ulnar	MC, LT	C8, T1	Abduct index finger
Infraspinatus	Suprascapular	UT	C5-6	Arm external rotation
Pronator teres	Median	LC, UT, MT	C6-7	Pronate forearm
Serratus anterior	Long thoracic		C5-7	Stabilize scapula
Supraspinatus	Suprascapular	UT	C5-6	Abduct arm
Triceps	Radial	PC, MT, LT	C7	Extend arm
Leg				
Extensor digitorum brevis	Peroneal	SP	L5-S1	Extend 2nd–4th toes
Gluteus medius	Superior gluteal		L4-S1	Abduct and medially rotate thigh
Medial gastrocnemius	Tibial	SP	S1-2	Plantar flex ankle
Rectus femoris	Femoral	LP	L2-4	Extend knee
Peroneus longus	Peroneal	SP	L5-S2	Eversion of foot
Biceps femoris, short head	Sciatic, peroneal division	SP	L5-S1	Flex knee
Tibialis anterior	Peroneal	SP	L4-5	Dorsiflex foot
Vastus medialis	Femoral	LP	L2-4	Extend knee

MC = medial cord; LC = lateral cord; PC = posterior cord; UT = upper trunk; MT = middle trunk; LT = lower trunk; SP = sacral plexus; LP = lumbar plexus.

9 Approach to Clinical Questions

No test is worthwhile in the absence of clinical information. All electrophysiologic tests are merely extensions of the physical examination. Failure to focus on the relevant history and examination is an open invitation to misdiagnosis. A good bat in the hands of a poor batter will not likely result in an acceptable hitting record, and a bat stuck in the ground is even less likely. Do a brief history and a focused examination at some point in the testing process and definitely prior to any EMG.

History should focus on the pattern of symptoms and their development and progression in time. Is there any numbness, weakness or pain? Where is it and how severe, and what are exacerbating and alleviating factors? Examination should include testing of sensory and motor function as well as reflexes. Armed with the clinical history and examination findings enables us to tailor the testing to the individual patient's needs. The worst complication of electrophysiologic testing is failure to get the needed information, or worse, to muddy the water with misinformation. The first task of neuromuscular diagnosis is to correctly classify the disorder.

Classification and Identification of Neuromuscular Disorders

Nerve conduction studies (NCS) and EMG classify disorders according to pathophysiology. With additional information on timing and chronicity of the disorder, the differential diagnosis can be substantially narrowed. Table 9.1 shows the findings with different pathologies.

The most common finding among patients with generalized neuromuscular disorders is axonal neuropathy. This has a huge differential diagnosis and a distinct diagnosis is not established in all patients.

Recommended Evaluation for Common Clinical Presentations

Physicians should supervise the performance of NCS and should perform EMGs themselves. A prominent reason is because the NCS and EMG is an expert consultation. Table 9.2 shows recommended evaluation for patients with selected clinical presentations. Table 9.3 shows neurophysiologic findings in patients with some common clinical disorders.

TABLE 9.1 Clinical correlations of nerve conduction studies and electromyography

Disorder	Nerve conduction studies	Electromyography
Axonal neuropathy	Normal except reduced CMAP amplitude	Acute and/or chronic denervation
Demyelinating neuropathy	Slow motor and sensory NCVs, prolonged or absent F waves	Reduced recruitment
Motoneuron degeneration	Normal except perhaps reduced CMAP amplitude	Acute and chronic denervation
Sensory neuron degeneration	Normal except reduced SNAP amplitude	Normal
Myopathy	Normal except occasional low CMAP amplitude	Myopathic MUPs and early recruitment
Neuromuscular transmission defect	Normal or reduced CMAP amplitude. Abnormal repetitive stimulation or response to paired stimulation	Normal or mild denervation

CMAP=compound motor action potential; NCV=nerve conduction velocity; SNAP=sensory nerve action potential; MUP=motor unit potential.

TABLE 9.2 Recommended evaluations for selected clinical presentations

Problem	Nerve conduction studies	Electromyography
Brachial plexopathy	Median motor NCV Median sensory NCV Median F wave Ulnar motor NCV Ulnar sensory NCV Ulnar F wave	First dorsal interosseus, pronator teres, biceps, triceps, deltoid, supraspinatus
Carpal tunnel syndrome	Median motor NCV Median sensory NCV	Abductor pollicis brevis
Cervical radiculopathy	Median motor NCV Median sensory NCV Median F wave Ulnar motor NCV Ulnar sensory NCV Ulnar F wave	Abductor pollicis brevis First dorsal interosseus Extensor digitorum communis Biceps Triceps Deltoid Cervical paraspinals
Critical illness polyneuropathy vs. myopathy	Tibial motor NCV Tibial F wave Peroneal motor NCV Sural sensory NCV	Tibialis anterior Medial gastrocnemius Vastus medialis Rectus femoris First dorsal interosseus Biceps Deltoid
Dysarthria or dysphagia	Median motor NCV Median sensory NCV Median F wave	Tongue Deltoid First dorsal interosseus Rectus femoris Tibialis anterior
Fatigue or weakness that is not localized	Peroneal motor NCV Peroneal sensory NCV	First dorsal interosseus Biceps

(Continued)

TABLE 9.2 (Continued)

Problem	Nerve conduction studies	Electromyography
	Consider repetitive stimulation	Deltoid Rectus femoris Vastus medialis Tibialis anterior
Foot drop	Peroneal motor NCV Peroneal F wave Tibial motor NCV	Tibialis anterior Peroneus longus Biceps femoris, short head
Guillain–Barré syndrome	Median motor NCV Median sensory NCV Median F wave Peroneal motor NCV Peroneal F wave Sural sensory NCV Ulnar motor NCV Ulnar sensory NCV Ulnar F wave	First dorsal interosseus Tibialis anterior Cervical paraspinals Rectus femoris
Hand pain, weakness, or both	Median motor NCV Median sensory NCV Ulnar motor NCV Ulnar sensory NCV	First dorsal interosseus Abductor pollicis brevis Abductor digiti minimi
Lumbar plexopathy	Tibial motor NCV Peroneal motor NCV Sural sensory NCV Tibial F wave Peroneal F wave	Tibialis anterior Medial gastrocnemius Vastus medialis Rectus femoris
Lumbar radiculopathy	Tibial motor NCV Sural sensory NCV Tibial F wave H reflex Peroneal motor NCV	Tibialis anterior Medial gastrocnemius Vastus medialis Lumbar paraspinals
Meralgia paresthetica	Tibial motor NCV Sural sensory NCV Femoral distal latency Lateral femoral cutaneous nerve conduction	Rectus femoris Vastus medialis Tibialis anterior Medial gastrocnemius
Motoneuron disease	Tibial motor NCV Sural sensory NCV Medial motor NCV Medial sensory NCV	Muscles in three nerve distributions in three extremities
Peripheral neuropathy	Tibial motor NCV Sural sensory NCV Tibial F wave Study at least two extremities	Tibialis anterior Medial gastrocnemius
Peroneal palsy	Peroneal motor NCV Superficial peroneal sensory NCV Peroneal F wave	Tibialis anterior Peroneus longus Short head of biceps femoris
Radial neuropathy	Radial motor NCV Radial sensory NCV	Extensor digitorum communis Triceps
Sciatic neuropathy	Tibial motor NCV Peroneal motor NCV Sural sensory NCV Tibial F wave Peroneal F wave	Tibialis anterior Medial gastrocnemius Vastus medialis Gluteus medius

(Continued)

TABLE 9.2 (Continued)

Problem	Nerve conduction studies	Electromyography
Spinal stenosis	See Lumbar radiculopathy	See lumbar radiculopathy
Tarsal tunnel syndrome	Medial plantar motor NCV Lateral plantar motor NCV Tibial motor NCV Sural sensory NCV	Abductor hallucis
Thoracic outlet syndrome	Medial motor NCV Medial sensory NCV Ulnar motor NCV Ulnar F wave Ulnar sensory NCV	Abductor pollicis brevis First dorsal interosseus Extensor digitorum communis
Ulnar neuropathy	Ulnar motor NCV across and below elbow Ulnar sensory NCV	First dorsal interosseus Abductor digiti minimi

NCV = nerve conduction velocity.

TABLE 9.3 Findings with common clinical disorders

Disorder	Nerve conduction studies	Electromyography
Amyotrophic lateral sclerosis	Decreased CMAP amplitude or normal NCS	Widespread active and chronic denervation
Carpal tunnel syndrome	Increased motor DL Increased median sensory DL	Denervation in APB
Guillain–Barré syndrome	Increased F-wave latency Decreased motor and sensory NCV	Usually normal Reduced recruitment Occasional mild denervation
Muscular dystrophy	Decreased CMAP amplitude Otherwise normal NCS	Myopathic features primarily in proximal muscles
Myasthenic gravis	Abnormal decrement to repetitive stimulation	Usually normal; may be fibrillation potentials
Peroneal palsy	Decreased peroneal motor NCV Decreased peroneal CMAP amplitude	Denervation in TA and EDC Biceps femoris short head is normal
Polymyositis	May be decreased CMAP amplitude Otherwise normal NCS	Myopathic features in mainly proximal muscles
Pronator teres syndrome	Decreased median motor NCV	Denervation of APB and FDS Normal pronator teres
Radiculopathy	Usually normal NCS Normal F waves	Denervation in a dermatomal distribution
Thoracic outlet syndrome	Decreased ulnar SNAP amplitude Increased ulnar F wave latency	Denervation of hand intrinsics including APB
Tarsal tunnel syndrome	Increased tibial motor DL Decreased NCV of medial and/or lateral plantar nerves	Normal or denervation of abductor hallucis brevis
Ulnar neuropathy, Guyon's canal	Increased ulnar motor DL Decreased ulnar sensory NCV	Denervation of first-dorsal interosseus

(Continued)

TABLE 9.3 (Continued)

Disorder	Nerve conduction studies	Electromyography
Ulnar neuropathy, cubital tunnel	Slow conduction across cubital tunnel	Denervation of first-dorsal interosseus, ADM, and ulnar-innervated portion of FDP

CMAP = compound motor action potential; NCS = nerve conduction studies; DL = distal latency; APB = abductor pollicis brevis; NCV = nerve conduction velocity; TA = tibialis anterior; EDC = extensor digitorum communis; FDS = flexor digitorum superficialis; SNAP = sensory nerve action potential; ADM = abductor digiti minimi; FDP = flexor digitorum profundis.

10 Neuropathies

Peripheral Neuropathy

The approach to this particular question should include sensory and motor nerve conduction studies (NCS) in at least one arm and one leg. Motor and sensory nerves in each limb should be tested, preferably in at least two segments, a more proximal one as well as a distal one. EMG should be done in at least one proximal and one distal muscle in each limb. This allows an opportunity to gauge the extent of the axonal involvement and also to look for conduction block and any degree of demyelination. In the western hemisphere the most common cause of neuropathy is diabetes mellitus. Diabetes can cause a wide range of abnormalities including axonal neuropathy, demyelinating neuropathy, and a mixed picture. Electromyography is sensitive and helpful in identifying acute and chronic denervation changes associated with axonal involvement in neuropathy. It should be considered an essential part of a complete electrophysiologic assessment. Table 10.1 contains a list of some of the commonly seen peripheral neuropathies and their expected electrophysiologic findings.

Generalized neuropathies can be primarily axonal, primarily demyelinating, or a combination of both.

Important Neuropathies

The most common neuropathies encountered in routine clinical practice are:

- Diabetic peripheral polyneuropathy
- Carpal tunnel syndrome
- Ulnar neuropathy at the elbow or wrist
- Idiopathic peripheral polyneuropathy
- Peroneal neuropathy
- Critical illness polyneuropathy
- Amyotrophic lateral sclerosis

TABLE 10.1　Important neuronopathies

Neuronopathy	Clinical features	Nerve conduction studies and electromyography
Amyotrophic lateral sclerosis	Weakness and incoordination of limbs Dysarthria and dysphagia are common Corticospinal tract signs	Widespread active and chronic denervation Fasciculations Proximal and distal muscles are affected
Spinal muscular atrophy	Weakness most prominent proximally and of the legs	Active and chronic denervation, mainly proximally Fasciculations
Primary lateral sclerosis	Leg weakness and corticospinal tract signs	Normal
Poliomyelitis	Focal or multifocal weakness, acutely	Chronic denervation Normal NCS

These fall into three groups:

- Neuronal degenerations
- Polyneuropathies
- Mononeuropathies

Neuronal Degeneration

Amyotrophic Lateral Sclerosis

Amyotrophic lateral sclerosis (ALS) is the most common neuronal degeneration affecting the peripheral nervous system. There is degeneration of upper and lower motoneurons without involvement of sensory systems. Motor and sensory nerve conduction velocities (NCVs) are normal, although the compound motor action potential (CMAP) may be of reduced amplitude. EMG shows widespread denervation. Examination for motoneuron disease including ALS should include examination of at least three muscles in different nerve distributions in each of three limbs. The head may be substituted for one limb, and tongue examination is quite easy. Electrical signs of acute and chronic denervation are seen, but acute signs may be subtle, depending on the stage of the disease. Fasciculations are common in ALS, but by themselves are not signs of denervation, since benign fasciculations may occur.

Spinal Muscular Atrophy

Spinal muscular atrophy (SMA) is a neurodegenerative condition with only lower motoneuron involvement. SMA can affect young children as well as adults. Motor NCVs are normal or only mildly slowed. Sensory NCVs are normal; EMG shows prominent denervation with fibrillation potentials and positive sharp waves. Recruitment may be incomplete, with rapid firing of single units. Long-duration polyphasic motor unit potentials (MUPs) are rare early in the disease but then become prominent with ongoing motor unit reorganization.

Poliomyelitis

Poliomyelitis is caused by a neurotropic enterovirus that destroys anterior horn cells. Polio is rare today, and we have personally only identified one case, related to vaccine. Motor NCVs are normal or near normal. EMG shows chronic denervation in survivors. Since polio is focal or multifocal, the EMG findings are most prominent in clinically affected muscles. When polio affects patients at a young age, the degree of polyphasia is less but the MUPs are larger, indicating better reinnervation of denervated muscle fibers. Signs of motor unit reorganization are often seen in muscles that were clinically unaffected.

Postpolio syndrome has received tremendous attention from the lay and medical communities. Patients complain of a variety of symptoms, including increasing weakness and pain. EMG shows signs of chronic denervation with high-amplitude polyphasic MUPs, with some occasional fibrillation potentials. The clinician must determine whether there are any signs of new disease responsible for the symptoms. The neurophysiologist can, in most instances, document the prior denervation on the basis of EMG alone. Prominent signs of active denervation should suggest new axonal damage. Repetitive stimulation and single-fiber EMG are of little benefit for evaluation of weakness in patients with postpolio syndrome, since polio will produce abnormalities.

Polyneuropathies

Generalized peripheral neuropathies are commonly seen in the course of electrodiagnostic work. Abnormalities in NCV and amplitude in a general sense help to indicate the amount of demyelination and axonal involvement respectively. Nerves primarily consist of two components: the axon and the myelin sheath. Disease processes may affect one or the other or both, and a well-performed NCS can be of great aid to the clinician in determining the presence and severity of a wide array of conditions. Primarily axonal neuropathies may affect sensory and/or motor nerves and tend to produce decreased SNAP and CMAP amplitudes. The absolute nerve conduction velocity may be normal or only slightly slowed, although severe axonal neuropathies eventually produce some degree of secondary demyelination. Primarily demyelinating neuropathies produce slowing of the maximal NCV that may be mild or marked. At its most severe demyelination can cause conduction block, which is a total failure of nerve impulse conduction past a certain point. Clinically this is manifested by weakness in the affected muscle. Electrophysiologically this is manifested by a greater than 50% drop in CMAP amplitude between distal and proximal stimulation sites. Many neuropathies, such as those seen in diabetes mellitus and uremia, cause varying degrees of axonal loss and demyelination. Table 10.2 shows the features of some important polyneuropathies.

TABLE 10.2   Important polyneuropathies

Polyneuropathy	Clinical features	Nerve conduction studies and electromyography
Diabetic sensorimotor neuropathy	Numbness and weakness distally Often with burning feet	NCS may be normal but more commonly mildly slow Chronic denervation
Diabetic mononeuropathy multiplex	Common mononeuropathies are ulnar, peroneal, femoral	Multifocal slowing and denervation, especially ulnar, peroneal, femoral nerves
Hereditary motor sensory neuropathy type I (Charcot–Marie–Tooth)	Distal weakness and numbness Decreased reflexes AD inheritance	Slow NCVs, increased DLs, F waves delayed or absent
Hereditary motor-sensory neuropathy type II (Neuronal form of CMT)	Distal weakness with less suppression of DTRs than in type I AD inheritance	NCVs normal or mildly slowed Motor DLs increased CMAP and SNAP may be decreased
Hereditary motor-sensory neuropathy type III (Dejerine–Sottas)	AR inheritance Distal weakness and numbness	Slow NCVs Denervation is less prominent
Hereditary motor sensory neuropathy type IV (Refsum's disease)	CNS signs as well as peripheral neuropathy Upper motoneuron signs	Slow NCVs Denervation is less prominent

AD = autosomal dominant; AR = autosomal recessive; DTR = deep tendon reflexes.

Diabetic Neuropathy

Diabetes causes four basic types of neuropathy:

- Small-fiber polyneuropathy
- Large-fiber polyneuropathy
- Autonomic neuropathy
- Mononeuropathy and mononeuropathy multiplex

The small-fiber neuropathy is a distal, predominantly sensory neuropathy. C-fiber dysfunction predominates, although axonal degeneration also involves motor fibers. NCS may be normal because SNAPs are dominated by large-diameter fibers. EMG is usually normal, as well. The most sensitive test of small-fiber dysfunction is the sympathetic skin response.

The large-fiber neuropathy involves both motor and sensory axons, with particular loss of motor fibers, and vibration, joint perception, and two-point discrimination.

Autonomic neuropathy is also a predominant small-fiber disorder and is a common feature of almost all diabetic neuropathies.

Mononeuropathy can affect one or several nerves (mononeuropathy multiplex). Mononeuropathy multiplex is diagnosed by multifocal NCS changes, EMG abnormalities, or both. The most commonly affected nerves are femoral, lumbosacral plexus, oculomotor, abducens, ulnar, and median. Any cranial nerve can be affected. Mononeuropathy of the femoral nerve or proximal lumbar plexus is termed *diabetic amyotrophy*. Mononeuropathy affecting individual nerve roots is termed *diabetic radiculopathy*.

NCV is slowed in most patients with diabetic neuropathy even though the demyelination is often secondary to axonal degeneration. The EMG usually shows denervation in clinically affected muscles. Patients with isolated mononeuropathies or radiculopathies superimposed on a polyneuropathy will often have fibrillation potentials and neuropathic MUPs in clinically affected muscles.

Acute Inflammatory Demyelinating Polyradiculoneuropathy (Guillain–Barré Syndrome)

Acute inflammatory demyelinating polyradiculoneuropathy (AIDP) produces subacute weakness often with sensory changes. Areflexia is prominent and is a clue to diagnosis. Motor and sensory NCSs may be normal early in the course, but are clearly abnormal later, with slowing of proximal and distal conduction. The earliest changes are prolongation or absence of F waves and dispersion of the CMAP. EMG evidence of denervation is unusual but may develop in severe cases. Reduced recruitment and conduction block are seen early in the course.

When demyelination is severe, permanent axonal loss may occur in distal muscles. EMG and NCV usually return to normal, but conduction slowing may persist in severe cases.

There is an axonal form of AIDP, which might seem like an oxymoron. However, the clinical presentation can be very similar, though some reflexes may be preserved.

Chronic Inflammatory Demyelinating Polyneuropathy

Chronic inflammatory demyelinating polyneuropathy (CIDP) presents with weakness that may have a relapsing course. Motor and sensory NCVs are often slow, but the first finding is prolongation of F-wave latency. EMG is initially normal, but shows denervation with chronicity of disease. Active and chronic denervation are seen. With treatment, patients clinically improve and NCVs improve, but usually remain slow.

Some clinicians follow serial NCVs to assess response of CIDP to treatment. This is likely to be more routine with future practice.

Multifocal Motor Neuropathy

Multifocal motor neuropathy (MMN) presents with weakness and wasting without sensory findings. The suspected diagnoses at the time of evaluation are often ALS or progressive muscular atrophy. MMN produces segmental conduction block of motor axons with sensory axons being unaffected. F waves are absent or prolonged. Motor NCVs show conduction delay in isolated segments of the nerves. Sensory NCS is normal even through the segments with delayed motor conduction.

Other Axonal Neuropathies

There are a large number of axonal neuropathies, and many of these are toxic–metabolic. Most toxic neuropathies are due to axonal degeneration but demyelination is occasionally seen. Important agents that cause neuropathy include:

- Vincristine
- Cisplatin
- Lead
- Ethanol

There are no specific findings on NCS and EMG to differentiate between toxic axonal neuropathies and other causes of axonal neuropathy.

Hereditary Neuropathies

Hereditary Motor-Sensory Neuropathy Type I

Hereditary motor-sensory neuropathy type I (HMSN-I) is the demyelinating or hypertrophic form of Charcot–Marie–Tooth (CMT) disease. Motor and sensory NCVs are slow, often at 20–30 m/sec, and distal motor latencies are prolonged. F waves are absent or delayed. The EMG may suggest mild denervation, but the features of axonal degeneration are much less prominent than the features of demyelination. Slowing of nerve conduction may involve the facial nerve, causing an abnormal blink reflex.

Hereditary Motor-Sensory Neuropathy Type II

Hereditary motor-sensory neuropathy type II (HMSN-II) is the neuronal form of CMT. It is clinically similar to HMSN-I; however, the NCS and EMG findings are very different. Motor and sensory NCVs are either normal or mildly slow, indicating relative preservation of myelin. Motor distal latency may be increased, and CMAP and SNAP amplitudes may be reduced. The EMG shows signs of acute and chronic denervation, including prominent fibrillation potentials, fasciculations, and high-amplitude long-duration polyphasic MUPs.

Hereditary Motor-Sensory Neuropathy Type III

Hereditary motor-sensory neuropathy type III (HMSN-III), also known as Dejerine–Sottas disease is a demyelinating condition clinically similar to HMSN-I but distinguished by inheritance and severity of disease. Motor and sensory conduction velocities are slow and motor distal latencies are increased because of demyelination of distal nerve segments. Denervation changes on EMG are not as prominent as are the slowed NCVs.

A focal form of HMSN-III has been described. Only one nerve is affected. NCVs in other nerves are normal. The EMG is normal except in the distribution of the affected nerve, where there is prominent acute and chronic denervation.

Hereditary Motor-Sensory Neuropathy Type IV

Hereditary motor-sensory neuropathy type IV (HMSN-IV) is Refsum's disease, with both central and peripheral nervous system involvement. In the PNS, demyelination causes slowed NCVs. Degeneration of anterior horn cells causes EMG features of acute and chronic denervation.

Hereditary Motor-Sensory Neuropathy Type V

Hereditary motor-sensory neuropathy type V (HMSN-V) is a combination of peripheral neuropathy and upper motoneuron degeneration. The electrophysiologic changes are similar to HMSN-IV, and the disorders are distinguished by clinical features.

Friedreich's Ataxia

Friedreich's ataxia has autosomal recessive inheritance. Clinical and electrophysiologic signs of neuropathy are obvious. Motor NCS is usually normal, but SNAPs are absent or of low amplitude. If a SNAP is obtainable, sensory NCV is found to be normal or only mildly slowed because the sensory deficit is due to sensory neuron degeneration.

Mononeuropathies

Entrapment Neuropathies

Entrapment neuropathy is extremely common in clinical practice. While almost any nerve can suffer entrapment at one time or another, some are seen more commonly than others and we will confine our discussion to those most likely to be encountered in the typical electrophysiology laboratory. Table 10.3 shows the neurophysiologic features of some important mononeuropathies.

Median Nerve

Carpal Tunnel Syndrome   The most common entrapment neuropathy seen in our laboratory is carpal tunnel syndrome (CTS), which is a distal median neuropathy within the carpal tunnel at the wrist. Clinical diagnosis is usually straightforward, and the characteristic electrophysiologic findings no less so. The most common finding is slowing of distal sensory NCV and/or prolonged median nerve terminal latency, in the absence of abnormalities in other tested nerves. It is commonly bilateral and usually worse in the dominant hand. Denervation in the abductor pollicis brevis (APB) is sometimes seen in severe cases and warrants more aggressive treatment, typically consisting of surgical decompression of the nerve in the carpal tunnel. If denervation is seen in the APB or other distal median-innervated muscle, study of a distal ulnar-innervated muscle should be performed, such as the first dorsal interosseus (1st-DI), to rule out lower cervical radiculopathy. Also, if the distal median-innervated muscles show degeneration, proximal median-innervated muscles should be examined to look for median nerve proximal to the carpal tunnel.

Routine median motor and sensory conduction study may miss patients with clinically significant CTS. While at least 90% of patients with CTS will have some abnormality, this leaves many patients without objective proof of their lesion. To improve the detection of CTS, palmar stimulation and comparative studies are often performed. Palmar stimulation consists of stimulating the median nerve at 1 cm increments on either side of the palmar crease while recording the median SNAP. Patients with CTS will usually have

TABLE 10.3 Important mononeuropathies

Mononeuropathy	Clinical features	Nerve conduction studies and electromyography
Carpal tunnel syndrome	Pain and sensory change on the thenar side of the palm May be weakness of APB	Increased median motor DL Slow median sensory NCV through the carpal tunnel May be denervation in APB
Radial neuropathy at the spiral groove ("Saturday night palsy")	Wrist drop and perhaps triceps weakness, as well	Increased radial DL and slowed conduction through and below the lesion Denervation on EMG with chronic lesions
Ulnar entrapment at Guyon's canal	Sensory loss on ulnar side of the hand Weakness of ulnar intrinsic muscles Ulnar-innervated long finger flexors are unaffected	Slow ulnar motor and sensory NCVs at the wrist Denervation of the 1st-DI and ADM, with chronic lesions
Ulnar entrapment at the cubital tunnel	Weakness of ulnar intrinsics as well as ulnar-innervated long finger flexors	Slowed ulnar NCVs across the elbow region Ulnar CMAP and SNAP may be small with severe lesions
Peroneal entrapment at the fibular neck	Weakness of tibialis anterior with preservation of tibialis posterior function	Slow peroneal motor NCV across the fibular neck Denervation in the TA with chronic lesions Biceps femoris short head is spared
Femoral neuropathy	Weakness of knee extension with decreased knee reflex	Inconsistent femoral conductions are often not reliable Denervation in quadriceps with chronic lesions
Facial nerve palsy ("Bell's palsy")	Unilateral face weakness affecting upper and lower facial muscles No sensory loss	Abnormal direct response on blink reflex testing Denervation of affected muscles, with normal muscles elsewhere

DL= distal latency; APB = abductor pollicis brevis; NCV = nerve conduction velocity; 1st-DI = first dorsal interosseus; ADM = abductor digiti minimi; CMAP = compound motor action potential; SNAP = sensory nerve action potential; TA = tibialis anterior.

slowed conduction across one or more of these small segments. When performing incremental stimulation, it is important to not use excessive voltages or stimulus durations, otherwise the site of nerve activation may be far from the stimulation site, thereby compromising the effect of the technique.

Comparative studies include median–radial SNAP comparisons and median–ulnar SNAP comparisons. The former is preferred because of the lesser involvement in patients with DM. A difference in SNAP latency of 0.5 ms or greater is considered significant for interpretation as abnormal. Motor comparisons can also be helpful but less so than sensory comparisons. Difference in distal motor latency of the median and ulnar nerves should not exceed 1.8 ms; a greater difference suggests CTS. Comparison of median motor distal latencies (DLs) from the two sides should show a difference of no more than 1.0 ms; a greater difference indicates significant relative slowing on the side with the longer DL.

Pronator Teres Syndrome Pronator teres syndrome is due to compression of the median nerve as it passes through the pronator teres muscle in the proximal forearm. Median motor NCV through the forearm is slow, but the distal latency is normal. Sensory NCV is normal because the segment tested is distal to the pronator teres.

EMG shows active and chronic denervation of the APB, flexor digitorum superficialis, and median innervated portion of the flexor digitorum profundus. The pronator teres is not denervated because it is innervated proximal to the site of compression. This is an important feature that distinguishes the pronator teres syndrome from compression of the median nerve by the ligament of Struthers.

Compression by the Ligament of Struthers The ligament of Struthers is a fibrous band above the medial epicondyle. The clinical syndrome of median nerve compression by the ligament of Struthers resembles the pronator teres syndrome. Most median-innervated muscles of the forearm are weak. The differentiating feature is involvement of the pronator teres. In pronator teres syndrome, the pronator teres is tender but strength is normal, and there is no denervation on EMG. When median nerve compression occurs at the ligament of Struthers, the pronator teres is weak and shows acute and chronic denervation on EMG.

Anterior Interosseus Syndrome The anterior interosseus nerve is a branch of the median nerve that innervates several muscles in the forearm, including:

- Flexor digitorum profundus of the first two fingers
- Flexor pollicis longus
- Pronator quadratus

This syndrome is caused by injury to the nerve as it leaves the main trunk of the median nerve. NCVs are usually normal. Stimulation of the anterior interosseus nerve at the elbow may reveal increased distal latency of the CMAP recorded from the pronator quadratus. EMG shows denervation in the flexor pollicis longus, median-innervated portion of the flexor digitorum profundus, and pronator quadratus.

Ulnar Nerve

The ulnar nerve is susceptible to injury at the wrist, in the forearm, at the elbow, and above the elbow. In practice, the most common cause of ulnar neuropathy that we see is in diabetes, where ulnar nerve damage is prominent and stands out more severely than other components of neuropathy.

Entrapment at Guyon's Canal The ulnar nerve passes from the forearm into the hand through Guyon's canal. Ulnar compression at the wrist is analogous to CTS. Unlike compression of the ulnar nerve at the elbow, the ulnar-innervated flexors in the forearm are unaffected, and the sensory loss is confined to the ulnar side of the hand, sparing the forearm. When entrapment of the ulnar nerve is at the wrist, the dorsal side of the hand is clinically unaffected, and the palmar portion of the ulnar distribution near the wrist is unaffected, since these nerves separate from the ulnar nerve proximal to the wrist.

Ulnar NCVs are slowed across the wrist and the DL is increased when recording from the abductor digiti minimi (ADM). EMG shows denervation in the ADM and 1st-DI, but a normal pattern in the flexor digitorum profundus (ulnar side) and flexor carpi ulnaris.

Ulnar Entrapment at the Elbow The ulnar nerve is particularly susceptible to injury near the elbow. Compression is exacerbated by occupation or habit placing pressure on the ulnar groove or proximal forearm, obesity, propensity to acute elbow flexion, and polyneuropathies that predispose to pressure palsies.

NCS shows slowing of motor NCV across the elbow and slowing of distal sensory conduction. Occasionally, the lesion can be primarily axonal in nature and very little slowing is seen, but the EMG shows denervation in the ulnar-innervated hand intrinsic muscles. Denervation is not seen in the flexor carpi ulnaris because the nerve branch innervating this muscle leaves the ulnar nerve proximal to the elbow.

Lesion of the Palmar Branch of the Ulnar Nerve The palmar branch of the ulnar nerve innervates the ADM and provides sensation overlying digit 5. Therefore, lesion of this produces normal routine NCS including ulnar motor DL. However, DL to the 1st-DI is prolonged and denervation is seen in this muscle. The superficial sensory branch of the ulnar nerve may be damaged, especially in bicycle riders.

Radial Nerve

The most common sites of damage to the radial nerve are at the spiral groove in the arm and in the forearm as the nerve penetrates the supinator muscle. Injury can also occur with dislocation of the elbow and distally from compression by handcuffs.

Spiral Groove Radial nerve entrapment or damage in the spiral groove in the humerus is not rare and typically follows a night of overindulgence. This is the so-called Saturday night palsy. Wrist drop and finger drop are seen along with loss of sensation in the distribution of the radial sensory nerve. Abnormalities in radial sensory NCS are seen and in motor NCS. The amount of denervation in radial muscles seen on EMG is variable. This entity is easily distinguished from C7/C8 radiculopathy by demonstrating normal function in median and ulnar-innervated muscles.

Posterior Interosseus Syndrome As the radial nerve enters the forearm, it divides into the posterior interosseus nerve and a superficial sensory branch. The posterior interosseus nerve supplies the finger and wrist extensors. Injuries to the nerve cause weakness without sensory loss.

Radial NCV is slowed through the involved segment. EMG shows denervation in the wrist and finger extensors with sparing of the extensor carpi radialis longus and supinator. Both muscles are innervated by branches that arise proximal to the lesion.

Brachial Plexus Lesion

Upper Plexus Lesion (Erb's Palsy) Proximal muscles are more severely affected than distal muscles. Intrinsic muscles of the hand are spared. The most prominent denervation is in the deltoid, biceps, supraspinatus, and infraspinatus. The serratus anterior and rhomboids are spared because their innervation is proximal to the lesion.

The upper plexus is most commonly damaged by stretch, such as when the shoulder is forced down by impact or by pull downward of the arm. The upper plexus nerves are stretched and temporarily cease to conduct action potentials. If the stretch is sufficiently severe, there may be breakage of the axons, which can result in long-term impairment.

Radiation plexopathy is another important cause of upper plexus damage. There is less surrounding tissue around the upper plexus than the lower plexus, so radiation is more likely to have long-term deleterious effects.

Lower Plexus Lesion The lower plexus is susceptible to trauma, but is especially sensitive to neoplastic infiltration. Tumors arising from the apex of the lung will produce infiltration, compression of the lower plexus, or both. Denervation is most prominent in muscle innervated by the median and ulnar nerves (roots C8-T1).

Neoplastic infiltration is differentiated from radiation plexopathy mainly on clinical grounds, however, NCS and EMG can be helpful. Tumor infiltration usually slows conduction through the plexus, whereas conduction is usually normal in radiation plexopathy. EMG shows myokymia in many patients with radiation plexopathy but not in patients with neoplastic infiltration.

Brachial Plexitis NCVs may be normal or slow through the plexus. Severe lesions may cause slowing of distal median and ulnar conduction. The EMG may show denervation not only in weak muscles but also in muscles that seem clinically unaffected. Sufficient time must elapse before the EMG is abnormal, however. In practice, EMG shows denervation at about the time that the patient has developed significant weakness. Therefore, if there is pain but no weakness, the initial EMG may be normal and a later study may be more revealing.

Peroneal Neuropathy

Peroneal nerve entrapment at the fibular head, producing foot drop, is the most common entrapment syndrome of the lower extremity seen in our laboratory. This is commonly seen in trauma, but in our practice, we see this more commonly as a complication of bed rest in hospitalized patients.

Slowing of peroneal NCV at the fibular head is the classic finding. Denervation in tibialis anterior and peroneus muscles is variably seen. Care must be exercised to exclude the possibility of L5 radiculopathy and sciatic neuropathy. In the case of the former, denervation and weakness in posterior tibialis and hamstring muscles is seen in addition to findings in muscles supplied by the common peroneal nerve. In sciatic neuropathy, an abnormal EMG in the biceps femoris short head is sufficient to establish a more proximal localization of injury.

Tibial Neuropathy
Tibial nerve entrapment behind the medial malleolus is called *tarsal tunnel syndrome*. This condition is relatively uncommon, but frequently considered by podiatrists and orthopedic surgeons ordering NCS and EMG. The motor DL of the medial plantar nerve, lateral plantar nerve, or both, are often increased. The medial plantar nerve is tested by stimulation of the tibial nerve with recording from the abductor hallucis. The lateral plantar nerve is tested by recording from the abductor digiti quinti. Sensory conduction through the tarsal tunnel is performed by stimulation of the medial or lateral plantar nerves and recording from the tibial nerve behind the medial malleolus. Because of the tremendous effect of foot temperature on nerve conduction, especially sensory, subtle differences in latency should be interpreted with caution, and the absence of a SNAP on the affected side should support the clinical diagnosis. Motor DLs are less dependent on temperature than sensory conduction. Differences in left–right and medial–lateral conductions are evidence of tarsal tunnel syndrome.

Some neurophysiologists use incremental stimulation through the tarsal tunnel, similar to that performed for CTS. This probably does not add anything to careful measurement of medial and lateral plantar motor DL and sensory latency, however.

Sciatic Neuropathy
Sciatic neuropathy is seen less commonly but often can masquerade as peroneal neuropathy. In most sciatic nerve injuries the peroneal division bears the brunt of the injury with relative sparing of the posterior tibial division. The reason for this is unclear. EMG revealing denervation in the distribution of both divisions of the sciatic nerve is helpful in establishing this diagnosis.

Piriformis Syndrome Piriformis syndrome is compression of the sciatic nerve by the piriformis muscle. It is a clinical diagnosis with no unique electrophysiologic findings. EMG shows denervation in the distribution of the peroneal division of the sciatic nerve. There is little or no involvement of the tibial division. Denervation of distal muscles with sparing of gluteus medius is supportive of piriformis syndrome, since this muscle receives innervation from a branch of the sciatic nerve that arises proximal to the lesion. Piriformis syndrome is differentiated from peroneal entrapment at the fibula by involvement of the short head of the biceps femoris; in peroneal entrapment, this muscle is unaffected, since innervation arises from the peroneal division proximal to the knee.

Sciatic Stretch The sciatic nerve can be damaged by stretching, which can occur especially in the lithotomy position–abduction, and slight flexion of the legs. This position stretches the sciatic nerve as it passes through the sciatic notch into the leg. The patient awakens from anesthesia with weakness of sciatic-innervated muscles. This syndrome can also be seen in patients who have sudden forward flexion at the waist.

NCVs of the tibial and sural nerves are usually normal. F-wave latency may be increased. EMG is normal or shows only a decreased number of MUPs immediately after the injury; electrodiagnostic study is often performed soon after the injury, before electrical signs of denervation develop.

Radiculopathy

The search for radiculopathy is a daily task for the electromyographer. Radiculopathy results from irritation of the nerve root as it exits the spinal cord. Most irritative lesions are extradural but the inflammatory process can occasionally extend somewhat into the subarachnoid space, giving rise to abnormalities in the cerebrospinal fluid. Needle EMG assessment is the cornerstone of diagnosis in these disorders. In choosing muscles for study, the clinical history and neurological examination is of paramount importance. Muscles within the suspected abnormal myotome should be selected, along with muscles from adjacent myotomes. As an example, for a suspected C7 radiculopathy, one could study the bicep, tricep, pronator teres, anconeus, first dorsal interosseus, and abductor pollicis brevis. Abnormalities found exclusively in C7-innervated muscles traveling in different peripheral nerves can clearly reveal the process as a radiculopathy as opposed to a mononeuropathy. When the clinical picture is vague and the examination normal, a representative sampling of muscles innervated by several levels is appropriate. An example for the upper extremity would be as follows: deltoid, supraspinatus (C5); bicep, brachialis (C6); tricep, pronator teres (C7); extensor indicis proprius, 1st-DI, abductor pollicis brevis (C8). Routine nerve conduction studies should always precede electromyography to eliminate confusion. If there is denervation in all tested C8 muscles for example, NCS can help determine whether this is due to a distal generalized neuropathy. Radiculopathy can often produce abnormalities of mild slowing and decreased amplitudes in the motor NCS with relatively normal sensory NCS. This pattern coupled with characteristic abnormalities in the EMG is helpful in determining the diagnosis of radiculopathy. Table 10.4 describes commonly seen radiculopathies and their respective EMG findings.

Mononeuropathy Multiplex

Diabetes mellitus and polyarteritis nodosa are the most common causes of multiple mononeuropathies in the United States. Leprosy is a common cause worldwide.

TABLE 10.4 Radiculopathies

Radiculopathy	Clinical findings	Nerve conduction studies and electromyography
C5	Sensory loss radial forearm Motor: deltoid, biceps Reflex: biceps	NCS usually normal Denervation: deltoid, biceps, paraspinals
C6	Sensory: digits 1 and 2 Motor: biceps, brachioradialis Reflex: biceps	NCS usually normal Denervation: biceps, paraspinals
C7	Sensory: digits 3 and 4 Motor: wrist extensors, triceps Reflex: triceps	NCS usually normal Denervation: EDC, triceps, paraspinals
C8	Sensory loss digit 5 Motor: intrinsic hand muscles Reflex: triceps	NCS usually normal Denervation: triceps, intrinsics, paraspinals
L2	Sensory: lateral anterior upper thigh Motor: psoas, quads Reflex: none	NCS usually normal Denervation: quads, paraspinals
L3	Sensory: lower medial thigh Motor: psoas, quads Reflex: knee	NCS usually normal Denervation: quads, paraspinals
L4	Sensory: medial lower leg Motor: TA, quads Reflex: knee	NCS usually normal Denervation: quads, paraspinals
L5	Sensory: lateral lower leg Motor: Peroneus longus, gluteus medius, TA, toe extension Reflex: none	NCS usually normal Denervation: TA, paraspinals
S1	Sensory: lateral foot, digit 4 and 5, outside of sole Motor: gastroc, gluteus maximus Reflex: ankle	NCS usually normal Denervation: gastroc, paraspinals

EDC = extensor digitorum communis; TA = tibialis anterior.

Polyarteritis Nodosa

Polyarteritis nodosa is an idiopathic connective tissue disorder characterized by multifocal vasculitis. This is a clinical diagnosis and can be supported but not confirmed by NCS and EMG.

Affected nerves show slowed or blocked conduction at the site of arteritis. EMG shows acute and chronic denervation in affected areas. Complete denervation causes fibrillation potentials and positive sharp waves and absence of MUPs with attempted voluntary effort.

Leprosy

Leprosy is probably one of the most common causes of neuropathy worldwide. The neuropathy is caused by either primary nerve infiltration or by infarction of the vasa nervorum.

NCS shows blocked or slowed conduction. EMG shows active and chronic denervation in muscles innervated by affected nerves. If the neuropathy is predominantly due to cutaneous vasculitis, only the distal sensory branches may be involved.

11 Myopathies

The major categories of muscle disease are:

- Dystrophies
- Inflammatory myopathies
- Metabolic myopathies

Myopathies are usually diagnosed by a presentation of muscle weakness without sensory loss, elevated CK in most disorders, and myopathic findings on EMG. Nerve conduction velocity (NCV) and EMG findings are typically:

- Normal motor and sensory NCVs
- Increased insertional activity
- Myopathic motor unit potentials
- Early recruitment

Motor and sensory conduction velocities are normal in most disorders, although selected metabolic disorders affect peripheral nerves in addition to muscles. Compound motor action potential (CMAP) amplitude may be reduced because muscle fibers fail to be activated. EMG is essential for diagnosis. Insertion may elicit complex repetitive discharges. At rest, there are fibrillation potentials and positive sharp waves. Motor unit potentials (MUPs) are reduced in amplitude and brief in duration. With increasing effort, units are recruited earlier than normal because of reduced tension output of the muscle fibers, termed *early recruitment*.

Muscular Dystrophies

Dystrophic Disorders: Duchenne and Becker's Muscular Dystrophies

Duchenne and Becker's muscular dystrophies are characterized by normal NCVs and myopathic findings on EMG. The EMG shows fibrillation potentials, complex repetitive discharges, and early recruitment. Fibrillation potentials are not as prominent as with inflammatory myopathies and with denervation. In later stages of the disease, muscle is replaced by fat and connective tissue, and insertional activity is reduced or absent. Table 11.1 summarizes some of the important muscular dystrophies.

Limb-Girdle Dystrophy

Limb-girdle dystrophy is the common phenotypic expression of several disorders. Many male patients diagnosed with limb-girdle dystrophy actually have Becker's dystrophy, a neuropathic disorder, or a metabolic myopathy. The EMG shows myopathic features in the majority of patients. Those with neuropathic features probably have a spinal muscular atrophy.

Myotonic Dystrophy

Myotonic dystrophy is characterized by myotonia on EMG. Myotonia is the repetitive discharge of muscle fibers with an initially high frequency that gradually declines. This produces a "dive bomber" sound in the audio monitor. Myopathic MUPs may also be seen. Occasional neuropathic features may

TABLE 11.1   Muscular dystrophies

Disorder	Clinical findings	Nerve conduction studies and electromyography
Duchenne type muscular dystrophy	Weakness Pseudohypertrophy of calves Gower's sign XR	Myopathic MUPs Early: increased insertional activity, fibs Late: reduced insertional activity, no fibs
Becker's muscular dystrophy	Features of Duchenne type muscular dystrophy with later onset and longer survival XR	Similar to Duchenne type muscular dystrophy with possibly more fibs CRDs
Scapuloperoneal dystrophy	Footdrop, shoulder weakness Onset in childhood AD or XR	Myopathic MUPs
Humeroperoneal dystrophy (Emery–Dreifuss)	Weakness of arms, shoulders, legs in peroneal distribution Contractures of neck, elbows Cardiac defects XR	Mixed myopathic and neuropathic patterns on EMG
Facioscapulohumeral dystrophy	Facial weakness followed by arm weakness and scapular winging AD	Early: mild changes Later: myopathic MUPs
Limb-girdle dystrophy	Progressive weakness of pelvic and shoulder girdle muscles in adults Sporadic, AR, or rarely AD	Usually myopathic MUPs Occasionally signs of denervation
Myotonic dystrophy	Muscle wasting and weakness with clinical myotonia Associated baldness, cataracts, endocrinopathies AD	Myotonia in patients and some asymptomatic relatives May be myopathic MUPs and mild slowing of NCV with reduced numbers of motor units

XR = X-linked recessive; MUP = motor unit potential; fibs = fibrillation potentials; CRD = complex repetitive discharge; AD = autosomal dominant; AR = autosomal recessive; NCV = nerve conduction velocity.

include slow motor conduction and a reduced number of functioning motor units.

Other Muscular Dystrophies

Facioscapulohumeral muscular dystrophy (FSH) is an autosomal dominant disorder characterized by progressive weakness of the face and shoulder girdle.

Scapuloperoneal dystrophy is probably a variant phenotype of the same genetic disorder. The EMG may be normal in mild or early cases. Typical findings are myopathic MUPs and early recruitment. Fibrillation potentials are occasionally seen but are not prominent. Neuropathic changes are seen both by electrophysiologic and histologic studies in some patients, suggesting that this is a neuropathic disorder rather than a dystrophy.

Inflammatory Myopathies

Electrophysiologic changes in inflammatory myopathies are all the same. NCVs are normal, although CMAP amplitude may be reduced. EMG findings include fibrillation potentials and positive sharp waves, myopathic MUPs, and complex repetitive discharges. Fibrillation potentials are more prevalent in inflammatory myopathies than in muscular dystrophies. Abnormalities are more prominent in clinically weak muscles. The EMG is normal in approximately 10% of patients with polymyositis. This may be due to sampling error or to periods of relative inactivity during the course of the disease. Table 11.2 summarizes some of the inflammatory myopathies.

TABLE 11.2 Inflammatory myopathies

Disorder	Clinical findings	Nerve conduction studies and electromyography
Polymyositis	Proximal weakness Increased CK Inflammatory infiltrates on muscle biopsy	Myopathic MUPs Fibs and PSWs CRDs May be inconsistent findings between muscles
Dermatomyositis	Symptoms of polymyositis with discoloration of eyelids Rash on dorsum of fingers, especially MCP and PIP joints	See polymyositis, above
Inclusion body myositis	Proximal and distal weakness in adults, commonly asymmetric CK mildly increased	See polymyositis, above May also have some neuropathic features
Viral myositis	Muscle pain Increased CK	See inclusion body myositis, above

CK = creatine kinase; MUP = motor unit potential; fibs = fibrillation potentials; PSW = positive sharp wave; CRD = complex repetitive discharge; MCP = metacarpal phalangeal; PIP = proximal interphalangeal.

Metabolic Myopathies

Mitochondrial disorders may present as neuropathy or myopathy. Therefore, NCV and EMG should both be done when a mitochondrial myopathy is suspected. Tibial motor NCV and sural sensory NCV are a sufficient screen. If the sural SNAP is absent with a normal tibial motor NCV, a sensory NCV in the arm or superficial peroneal sensory NCV should be performed.

Endocrine myopathies may be associated with the following disorders:

- Cushing's syndrome
- Addison's disease
- Thyrotoxicosis
- Hypothyroidism (rarely)
- Hyperparathyroidism (rarely)

NCVs are normal except in hypothyroidism, in which case they may be slowed. The EMG may show minor myopathic features in all of these disorders.

Steroid myopathy (Cushing's syndrome) is usually a clinical diagnosis. The EMG usually does not show myopathic features. The NCVs are also normal, although CMAP amplitude may be reduced. Biopsy can help to differentiate inflammatory myopathy from steroid myopathy in patients being treated for inflammatory myopathy, but is seldom necessary.

Carnitine palmityl transferase (CPT) deficiency usually manifests normal NCV and EMG findings. Patients with carnitine deficiency have myopathic

TABLE 11.3 Metabolic myopathies

Disorder	Clinical findings	Nerve conduction studies and electromyography
Mitochondrial myopathies	Weakness with other manifestations, e.g., ptosis, ophthalmoplegia, cardiac abnormalities Mild increase of CK concentration	Myopathic MUPs, but findings may be subtle or absent May have neurogenic appearance as well
Myotonia congenita	Myotonia with or without cramps Onset in youth or young adults AD, AR, or sporadic	Myotonia RNS may show decrement with increasing stimulation frequency Some patients have myopathic MUPs
Hypokalemic periodic paralysis	Episodic weakness provoked by rest after large carbohydrate meal	Normal between episodes Reduced MUPs and insertional activity during attack Reduced CMAP amplitude
Hyperkalemic periodic paralysis	Episodic weakness provoked by rest after exercise or cold	Between attacks: possible myopathic patterns During attack: myotonia, increased insertion, reduced MUPs to volition to stimulation

CK = creatine kinase; MUP = motor unit potential; AD = autosomal dominant; AR = autosomal recessive; RNS = repetitive nerve stimulation; CMAP = compound motor action potential.

findings with small-amplitude polyphasic MUPs. Fibrillation potentials are seen but are rare. Table 11.3 summarizes features of some important metabolic myopathies.

Syndromes of Continuous Muscle Fiber Activity

Disorders of increased muscle fiber activity are classified into the following categories depending on the site of the defect:

- Tetanus
- Stiff-man syndrome
- Schwartz–Jampel syndrome
- Neuromyotonia (Isaac's syndrome)

Tetanus

Tetanus is characterized by involuntary discharge of motor units. The toxin works at the spinal level, blocking postsynaptic inhibition, thereby increasing the excitability of the motoneurons. The EMG shows repetitive MUPs that are abolished by peripheral nerve or neuromuscular block. The discharges are attenuated during sleep and with general or spinal anesthesia.

Stiff-Man Syndrome

Stiff-man syndrome is not really a disorder of muscle; it is due to excessive motoneuron activation. The reason for the enhanced discharge is unknown. Excessive motor unit activation results in involuntary muscle contraction involving predominantly proximal muscles. Affected muscles show normal MUPs with coactivaion of agonists and antagonists. Discharges are attenuated by sleep, general anesthesia, benzodiazepines, peripheral nerve block, or neuromuscular block.

Schwartz–Jampel Syndrome

Schwartz–Jampel syndrome is characterized by multiple congenital anomalies in association with increased muscle fiber activity. The defect is probably at the nerve terminal. Clinically, the muscle activity looks similar to that of myotonia, but on EMG the discharges have the appearance of complex repetitive discharges, lacking the frequency modulation of true myotonia. The discharges are abolished by neuromuscular block but not by nerve block.

Neuromyotonia

Neuromyotonia is a term used to differentiate muscle fiber activity of nerve terminal origin from myotonia that is of muscle membrane origin. Neuromyotonia is repetitive activity of single muscle fibers rather than of complete motor units. The muscle fibers discharge repetitively at frequencies that are initially high and gradually decline. This is similar to true myotonia, but the discharges have an invariant decline in frequency rather than a waxing and waning frequency. Also, these discharges are apparent at rest, whereas myotonia is evoked primarily by needle insertion. The amplitude of MUPs may be reduced because of the loss of functioning muscle fibers due to continuous activity. The defect is probably in the terminal motor axon. Therefore, the discharges are abolished by neuromuscular block but not by peripheral nerve block, spinal block, or general anesthesia.

Other Disorders of Muscle

Myotonia Congenita

Myotonia congenita (Thomsen disease) is an autosomal dominant disorder characterized by involuntary contraction of muscles evoked especially by a forceful contraction. For example, a patient may grasp a railing then be unable to release the grip. Repeated contractions produce progressively less contraction.

NCVs are normal. EMG shows myotonia. Repetitive stimulation produces a decremental response that is greater with high frequencies of stimulation.

Paramyotonia Congenita

Paramyotonia congenita (of Eulenberg) produces weakness with myotonia that is exacerbated by repetitive activation. Potassium is increased, as in hyperkalemia periodic paralysis. Cold can evoke the weakness with myotonia, but with progressive chilling the myotonia disappears and the muscles are flaccid. Repeated contractions exacerbate the myotonia, in contrast to myotonia congenita, in which repeated contractions alleviate the myotonia.

NCVs are normal. Repetitive stimulation often results in a decremental response. EMG shows myotonia with needle insertion or percussion.

Periodic Paralysis

The periodic paralyses are a family of disorders all characterized by abnormal loss of excitability of the muscle fiber membrane. The common thread is the symptom of episodic muscle paralysis, however, the clinical presentations are quite diverse.

Familial hyperkalemic periodic paralysis produces episodes of weakness that are provoked by periods of rest after exercise, cold, and sleep. NCVs are normal. The EMG is often normal but may show some myopathic features. During an attack, motor nerve stimulation activates fewer muscle fibers. Therefore, the CMAP is smaller.

Myotonia may be seen in hyperkalemic periodic paralysis, blurring the distinction between this entity and paramyotonia congenita.

12 Neuromuscular Transmission Defects

Three principal conditions constitute 99% of all cases of neuromuscular junction (NMJ) transmission defects seen in the average EMG laboratory: myasthenia gravis, Lambert–Eaton myasthenic syndrome (LEMS), and botulism. Myasthenia gravis is a postsynaptic defect, and LEMS and botulism are presynaptic defects with distinct pathophysiology. Defects in NMJ function cause weakness of affected muscles with no sensory loss. In addition to abnormal results in repetitive nerve stimulation (RNS) testing, increased jitter in single-fiber EMG is seen with all NMJ disorders.

Clinical findings that would suggest a neuromuscular transmission disorder include weakness that changes with use, especially showing improvement. Deterioration with activity is common with many disorders producing muscle weakness. Ptosis and diplopia suggests myasthenia gravis, especially when there is generalized weakness, although ocular myasthenia has no systemic weakness. Myasthenia is especially likely when the ocular motor deficit cannot be explained by a single cranial nerve deficit.

Repetitive stimulation and single-fiber EMG are performed on patients who are being evaluated for the possibility of myasthenia gravis. Paired stimulation is performed mainly for botulism. LEMS is associated with abnormalities on both repetitive stimulation and single-fiber EMG testing.

Table 12.1 summarizes the clinical and neurophysiologic features of these disorders.

Myasthenia Gravis

Myasthenia gravis is a condition caused by production of IgG antibodies directed against the acetylcholine receptor on the postsynaptic membrane. This binding stimulates internalization and degradation of the receptor, therefore, there are fewer receptors available for binding with acetylcholine. When an action potential depolarizes the presynaptic membrane, the transmitter cannot activate sufficient receptors to evoke an action potential in the muscle fiber. The sarcolemmal depolarization is not sufficient.

The clinical features of the disease include fluctuating weakness, fatigable weakness, and a predilection for involvement of oculobulbar muscles. Proximal limb-girdle weakness is more common and more severe than distal

TABLE 12.1 Disorders of neuromuscular transmission

Disorder	Clinical features	Nerve conduction studies and electromyography
Myasthenia gravis	Weakness that worsens with activity Ptosis and diplopia	Normal CMAP RNS: decremental response with low rates Abnormal single-fiber EMG with increased jitter and blocking
Lambert–Eaton myasthenic syndrome	Generalized or proximal weakness Dry mouth, impotence, and/or other signs of autonomic dysfunction	RNS: decremental response at low rates, incremental response at high rates Facilitation with exercise
Botulism	Gastrointestinal distress followed by weakness with bulbar involvement	Low CMAP amplitude, increases with exercise RNS: little decrement at low rates, increment at high rates

CMAP = compound motor action potential; RNS = repetitive nerve stimulation.

weakness. Pure ocular myasthenia is common. Routine electrophysiologic studies can be unimpressive. The absolute compound motor action potential (CMAP) amplitude should be normal in myasthenia gravis. Sensory nerve conduction studies (NCS) are normal. There should be no significant abnormalities in routine nerve conduction velocity studies. EMG is usually normal, but may occasionally show myopathic motor unit potentials but no fibrillation potentials or positive sharp waves.

RNS testing is the most important electrophysiologic assessment. As with any neurological disorder, studying clinically affected muscles increases diagnostic yield. At least two muscles should be selected for study, and at least one of these should be a muscle innervated by a cranial nerve. The orbicularis oculi and trapezius muscles are excellent choices. Techniques for study of arm and leg muscles are available and can be useful to increase diagnostic yield, but typically require the use of a stabilizing board to reduce movement artifact. Stimulation at low rates usually around 3–5 Hz produce decremental amplitudes of CMAPs when the first and fifth responses are compared. A decrement of greater than 11% is considered abnormal. Figure 12.1 shows response to repetitive stimulation.

Lambert–Eaton Myasthenic Syndrome

LEMS is typically considered a paraneoplastic condition associated with small-cell lung cancer and certain other malignancies. It results from antibodies directed against the voltage-gated calcium channels on the presynaptic nerve terminal. It is clinically manifest by fatigable proximal weakness and areflexia, with fewer oculobulbar findings than typically seen in myasthenia gravis. On continued muscle activity, strength and reflexes improve dramatically as a result of calcium accumulation in the presynaptic nerve terminal.

Repetitive Stimulation at Low Rates

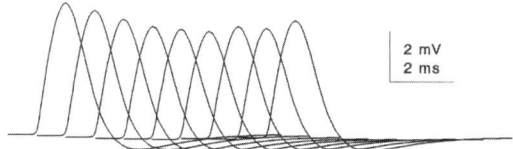

FIGURE 12.1 *Response to repetitive stimulation. The first response is the greatest in most instances, with a small fall-off in peak amplitude. The degree of decrement determines abnormality.*

Electrophysiologically, a decremental response to RNS is seen at low rates of stimulation and an incremental response is seen at high rates of stimulation of 50 Hz or more. The CMAP amplitude is almost always decreased, even in routine NCS. After brief exercise it can increase dramatically. This finding alone is strongly suggestive of LEMS. Sensory NCS are seldom affected. Choice of muscles for study is somewhat less critical than in myasthenia gravis because the defect in neuromuscular transmission seems to be more generalized, as opposed to the proximal predilection in myasthenia gravis.

Paired stimulation at short interstimulus intervals results in facilitation of the second response. This is similar to the response seen in botulism. In contrast, however, paired stimuli with interstimulus intervals of greater than 15 ms also results in facilitation of the second response. The pathophysiology of the facilitation with short and long latency paired stimuli is different. Therefore, this represents an actual facilitation of transmitter release.

Botulism

Botulism is a condition caused by a specific neurotoxin elaborated by several strains of Clostridium botulinum. The toxin consists of two peptide chains. The long chain is responsible for binding of the peptide to the cell surface and entry of the toxin. The short chain is the active neurotoxin. It is a serine metalloprotease that cleaves certain peptides in the presynaptic nerve terminal responsible for vesicle transport. This makes neurotransmitter release impossible, and permanently inactivates the synapse. The toxin is incredibly potent and it has been estimated that as little as four molecules of toxin are required to inactivate a single neuromuscular junction. A half liter of pure toxin would be sufficient to kill all people living on the earth.

Clinically, the syndrome of botulism is distinguished from the other NMJ disorders by signs of autonomic involvement, chief among which is paralytic ileus. Nausea, vomiting, constipation, and other gastrointestinal symptoms are common early in the course.

Electrophysiologically, findings are similar to those seen in LEMS with the exception that an incremental response at high rates of stimulation is not seen. EMG findings indicative of denervation also develop within a few weeks, which also help distinguish this NMJ disorder from the others. The NCS is normal except for reduced CMAP amplitude. Successive stimuli result in further reduction in CMAP amplitude. Paired stimuli can be helpful for diagnosis. At short intervals (less than 15 ms) the response to the second pulse is greater than to that of the first. This is because the second impulse activates some terminals that were not activated by the first pulse due to summation of end-plate potentials.

Neonatal Myasthenia

A classification of neonatal myasthenic syndromes was presented by Misulis and Fenichel (1989) based on pathophysiology (Table 12.2). Diagnosis of these syndromes depends on techniques not readily available in most laboratories, including:

- Muscle acetylcholinesterase assay
- In vitro intracellular electrophysiology
- Receptor binding studies

Most physicians can narrow the differential diagnosis by clinical diagnosis. Repetitive stimulation is not helpful, because it is abnormal in virtually all patients with genetic myasthenia. Normal repetitive stimulation responses at 3 Hz have been reported in some patients with impaired acetylcholine receptor function.

Acetylcholinesterase deficiency and slow-channel syndrome are characterized by repetitive discharges of muscle fibers by a single neural stimulus. A motor nerve is stimulated in the usual manner. A needle electrode is placed in the muscle for recording motor unit activity. In these disorders, a single stimulus produces prolonged depolarization of the postsynaptic membrane. In the case of acetylcholinesterase deficiency, the prolonged depolarization is

TABLE 12.2　Neonatal myasthenia

Genetic
Presynaptic
    Abnormal acetylcholine resynthesis or mobilization
    Abnormal acetylcholine release
Postsynaptic
    End-plate acetylcholinesterase deficiency
    Reduced number of acetylcholine receptors
    Impaired function of acetylcholine receptors
    Slow-channel syndrome

Acquired
Acetylcholine receptor antibody positive
    Transitory neonatal
    Juvenile myasthenia
        Generalized
        Mainly ocular
Acetylcholine receptor antibody negative
    Juvenile myasthenia
        Mainly ocular
        Relapsing ocular

due to failure of breakdown of acetylcholine and resultant sustained activation of postsynaptic receptors. In slow-channel syndrome, the sustained depolarization is due to prolonged open time of the ion channel associated with the acetylcholine receptor. Routine NCS and EMG are performed on all patients to look for neuropathies or myopathies that could be confused with genetic myasthenia. Transitory neonatal myasthenia occurs in children of myasthenic mothers. Repetitive stimulation at 3 Hz produces a decremental response.

Part IV Evoked Potentials

13   Evoked Potentials Basics

Evoked potentials (EPs) are the electrical responses to sensory stimulation. While there are motor EPs, they are not in routine clinical use and will not be discussed here. The three clinically used EPs are brainstem auditory (BAEP), visual (VEP), and somatosensory (SEP).

EPs are used mainly for the following clinical indications:

- Assessment for clinically silent lesions in possible multiple sclerosis (MS)
- Visual loss that might be due to optic neuritis
- Myelopathy when imaging studies do not show a structural abnormality
- Assessment of brainstem function in neonates

With the advent of magnetic resonance imaging (MRI), the role of EPs in neurologic localization has waned, somewhat; however, there is still some benefit to functional studies. One of the most persistent roles of EPs has been in evaluation of patients for possible MS. Patients present with paraparesis, hemiparesis, or visual loss, and may have MRI evidence of white-matter lesions. EPs can show functional abnormalities in affected and clinically unaffected regions. This documentation of multiple lesions supports the diagnosis of MS.

Visual loss usually triggers two types of evaluation—ophthalmological and MRI. Careful ophthalmological evaluation can determine whether the visual loss is prechiasmatic or postchiasmatic. Funduscopic changes may suggest optic neuritis, increased intracranial pressure, vascular abnormality, or primary retinal disorder. When the lesion is not felt to be ocular, MRI is commonly done to look for structural lesion. When none is found, EPs can document abnormality in optic nerve function that can suggest retrobulbar optic neuritis.

Paraparesis results in MRI or myelography, occasionally both. If no structural lesion is seen, the most likely diagnoses are transverse myelitis, MS, and anterior spinal artery infarction. Somatosensory EPs can show functional abnormalities in the spinal cord when the structural studies are normal. Anterior spinal artery syndrome usually results in normal EPs, but transverse myelitis and MS produces defective conduction through the involved segments.

These indications are for evaluation of clinical lesions without visualization on structural imaging. EPs are also helpful for assessment of clinically silent lesions in suspected MS. In this circumstance, the somatosensory EP is most valuable for patients with optic neuritis, and visual EP is most valuable for patients who present with paraparesis.

Physiology of Evoked Potentials

Neural Generators of Evoked Potentials

There are many generators of EPs, since responses are recorded from multiple sites along the afferent projection pathways. The VEP is most likely due to charge movement associated with conduction in projections from the lateral geniculate to the visual cortex. Central SEP activity is due to thalamocortical projections, but impulses conducted in the peripheral nerves and dorsal columns are recorded as well. The BAEP is recorded from the nerve volley in the eighth cranial nerve and potentials generated by tracts and nuclei in the brainstem. Specific locations of generators are discussed in the individual sections on VEP, BAEP, and SEP.

The generators of EPs are of two basic types: nerve bundles and nuclei. Nerve fiber bundles include both peripheral nerves and central tracts. The recorded potential is due to the advancing front of the compound action potential. The vector of this potential is determined by the direction of projection of the axons.

Normal and Abnormal Physiology

Potentials generated in nuclei are not easily described by vectors and axonal conduction. Movement of charge in nuclei is a combination of axonal action potentials and charge movement during synaptic transmission. Synapses are oriented in virtually all directions on a cell's dendrites and soma, such that it is impossible to predict the ultimate vector of positivity and negativity. Also, because of the complex orientation of the synapses, there is no guarantee that the field will conform to a simple dipole. Therefore, hypotheses of the sources of individual EP waves are developed on the basis of human pathology and animal studies in addition to a knowledge of basic neuroanatomy.

General Principles of Evoked Potentials

Evoked Potential Equipment

The equipment used to record EPs is similar to that used for routine EEG and EMG studies and should fulfill the guidelines for electrical safety outlined in Chapter 3. In general, modern and well-maintained machines meet the required limitations on allowable leakage current. Unsafe practices can endanger a patient even with the best equipment, however. The greatest risk is with SEPs, since the stimulus is an electrical pulse. The patient must be adequately grounded so that the path of current cannot traverse the heart or spinal cord. Normally, the path of current is between the two electrodes of the stimulator; however, current can flow from the stimulating electrode to ground if one lead has poor skin contact or high impedance. Minimal technical requirements for EP machines are tabulated in Table 13.1.

TABLE 13.1 Technical requirements for EPs

Parameter	Minimum value
Input amplitude range	5 µV to 50 µV
Input impedance of differential amplifier	10 Mohm
Common mode rejection ratio	10,000:1 (80 dB)
Time resolution	20 µsec/data point
Amplitude resolution	8 bit converter, 500 data points/trial
Averaging capacity	4,000 trials
Number of channels	For BAEP = 2 channels
	For SEP = 4 channels
Noise level of amplifier	2 µV root-mean-square
Chassis leak current	<100 µA

BAEP = brainstem auditory-evoked potential; SEP = somatosensory-evoked potential.

EP equipment ideally should be calibrated prior to each recording session; however, in practice, most modern machines perform a self-diagnostic test prior to the recording. Errors in function of a component of the system will usually be detected during boot-up of the machine or initial acquisition.

Acquisition of Signals

Stimulus and Response
Stimulus depends on the EP modality, and is covered in the respective chapters. The common thread is the stimulus–response circuit for each of the modalities. There are central projections for the sensory inputs and the central responses to those inputs produce the normal EP.

Averaging
EPs are of very small amplitude and in most instances cannot be seen without averaging. The EP is superimposed on EEG activity that is unrelated to the stimulus. In addition, muscle electrical activity and movement artifact contributes to the recorded potential.

Averaging brings out the EP by the assumption that most potentials not caused by the stimulus occur in a random fashion and will not produce a potential of substantial amplitude after averaging many trials. This assumption is generally true but has two possible sources of error. First, the stimulus may cause a slight movement of the patient that is sufficiently reproducible from trial to trial to be detectable in the average. An experienced neurophysiologist can usually identify such abnormal waveforms. Second, 60 Hz interference can appear to be a high-amplitude sinusoidal wave on averaging. To prevent this latter error, the stimulus rate should not be a harmonic of 60 Hz. The electronics of averaging and analog-to-digital conversion are discussed in detail in Chapter 2.

Artifact Rejection
Artifact rejection is an important part of noise reduction. An electronic window is created within specified times and voltages. Potentials larger than the

set voltage window are rejected from the average. If a trial contains a potential outside of this window, the entire trial is considered unreliable and therefore is rejected. When a trial is rejected, an indicator usually shows this on the screen. Also, at the end of the averaging period, most machines will display the number or percentage of rejected trials. High rejection rate is a strong indicator of poor technique or improper acquisition parameters.

Analysis Time
A time window is set during which the response is acquired. The sampling begins shortly after the stimulus. A brief delay is used to reduce stimulus artifact affecting the input amplifier. A high-voltage stimulus artifact may be conducted to the scalp electrodes, and subsequently overload the input amplifier. The high potential can alter the amplification of the amplifier for a brief time, a fraction of a second, but this may be time enough for the response to be missed. This is most important for SEPs, but there can also be artifact during BAEP from current movement in headphone leads. Pattern-reversal VEPs have effectively no stimulus artifact, though flash VEPs may produce potentials in anterior (usually reference) leads; this can be minimized by good electrode contact.

Limiting the duration of the recording was more important on earlier machines with low operating and storage capacities, but effectively is eliminated as a factor now. Nevertheless, there is no use to recording long after the response has passed. Modern EP equipment allows for changing of sweep speed, which in turn alters recording duration and sampling interval.

Replications
Two replications of each waveform are recommended for each EP. Consistency of waveforms is visualized if the traces are superimposed on the hard copy. Four replications may be necessary when recording SEPs to provide convincing identification of individual waves.

Display and Analysis of the Response

EP recordings are all bipolar. Up or down on the trace depends on which electrode is active and which is reference. For the VEP, this is clear, but it is not so clear for SEPs and BAEPs. For EEG, a negative potential delivered to the active electrode is shown as an upward deflection of the pen. Polarity conventions have recently become fairly consistent for EPs, and in general, tend to show the potentials of clinical importance as upward deflections. For SEPs and VEPs, traces are displayed so that a negative event at the active electrode produces an upward deflection. In contrast, for BAEPs, a positive event at the active electrode produces an upward deflection.

Normal and Abnormal Responses

Testing equipment and environment differ between laboratories, and each lab should set its own set of normative data, or at the very least, ensure that data obtained from normal subjects fit within the established published norms. For published norms to be used, not only must the data be verified in the lab but the techniques of acquisition must be identical.

There are maturational changes in EPs early in life, so the *Guidelines* (American Electroencephalographic Society, 1994) recommend that normative data be established for each week of the perinatal period, for each month of infancy, and each decade, thereafter. At least 20 subjects from each age group should be tested for each evoked response. Responses from the left and right sides of the same subject cannot be considered to be two subjects. Such data are used to establish normative data on interside differences in latency and amplitude.

Normal data are expressed as mean plus-or-minus standard deviation. A latency is considered to be abnormal if it exceeds 2.5–3.0 standard deviations from the mean. Some laboratories use two standard deviations; however, this allows an unacceptable percentage of false-positive interpretations. When normal data are displayed, it is evident that it does not fall into a "normal" distribution. Transformation of the data can result in a more normal distribution from which standard deviations can be calculated. Possible transformations include logarithm, square root, reciprocal, and so on.

Linear regression analysis allows a more precise evaluation of waveform latencies as a function of increasing age. The relationship between age and latency is not strictly linear; however, and such a level of precision is not needed.

Semmes-Murphey Clinic - Neurophysiology Laboratory
614 Skyline Drive, Jackson TN 38301 (731-423-1267)

Patient: Alpha Beta	*Test*: Visual Evoked Potential
Sex: Male	*Ordering MD*: Artz
Age: 81	*Testing MD*: Head
DOB: 08-13-20	*Tech*: ABC
ID: 1234567	*Test date*: 11-14-01
Height: 72 in	
Weight: 180 lbs	

History: Visual loss, possible optic neuritis.

Technical: Pattern reversal visual evoked potential study was performed using unilateral full-field stimulation.

Data:

	Left	Right	Interside comparison
P100 latency (ms)	97.6	134.0	36.4 difference
P100 amplitude (uV)	10.2	12.4	1.22 ratio

Interpretation: The absolute latency of the P100 from stimulation of the right eye is prolonged. The interside difference in P100 latency is prolonged. Interside amplitude ratio is normal.

Impression: Abnormal visual evoked potential study, consistent with a lesion in the right optic nerve.

Thomas C Head MD

FIGURE 13.1 *Sample normal evoked potential report.*

Reports

Reports should be concise but thorough. A sample report is shown in Figure 13.1. It is most helpful to put the data in tabular form. Highlighting abnormal values is also helpful. The interpretation can have two sections. The first describes what is abnormal, and the second gives the implications of the abnormalities. Interpretation of the data in light of the clinical history is essential, since many physicians ordering EPs are not experts in this field.

Hard copies of the waveforms should be kept with the patient's record in the laboratory. Most modern equipment allows for selected waves to be printed on the report along with the tabular data. This is helpful for other neurophysiologists but is not of interest for most clinicians.

14   Visual-Evoked Potentials

The visual-evoked potential (VEP) is the potential recorded from the occipital region in response to visual stimuli. The VEP differs from other evoked responses in that the response is a long-latency response. Brainstem auditory-evoked potentials (BAEPs) and somatosensory-evoked potentials (SEPs) are short-latency responses. There are long-latency components of the BAEP and SEP, but these are not routinely used for interpretation because they are too variable for clinical usefulness.

The VEP is the only evoked response that is visible without averaging. During routine EEG, photic stimulation is used to activate epileptiform discharges in patients suspected of having seizures. The photic stimulation is a bright flash delivered to subjects with their eyes closed. At flash frequencies of less than 5–7/sec, an evoked response is recorded from the occipital leads. A driving response is recorded at faster frequencies. The VEP is highly reproducible as long as the patient maintains fixation and has no change in visual acuity.

Methods

Stimulus

The VEP stimulus may be one of three types:

- Flash
- Full-field pattern reversal
- Half-field pattern reversal

Flash is used in patients who cannot cooperate with the level of fixation required for pattern-reversal stimulation. The latencies of the flash VEP are more variable than the pattern-reversal VEP, so the flash VEP can really only test continuity of the visual pathways. Full-field pattern reversal is the usual stimulus for the VEP. Each eye is examined individually, so the anterior visual pathways are evaluated especially well. Half-field pattern-reversal stimulation is used for localization of lesions behind the optic chiasm. Although many laboratories still perform half-field testing, modern imaging procedures have reduced the clinical application of this technique.

Flash

The strobe light is similar to that used for photic stimulation during conventional EEG. The strobe is placed in front of the patient's eyes, usually with the eyes closed. Plenty of light passes through the lids to activate the retina.

Presence of the flash VEP indicates continuity of the pathways from the retina to the lateral geniculate. Flash VEPs have been recorded in the absence of a functioning cortex. Therefore, flash stimuli are not used if reproducible waveforms can be obtained with pattern-reversal stimuli.

Pattern Reversal

A video display is placed in front of the patient, who is usually seated in a comfortable chair. The display is attached to the video output of the EP computer. The patient fixates on a small target in the center of the display. A black-and-white checkerboard pattern is produced that is unchanging between trials (Figure 14.1). When the trial begins, the black-and-white squares reverse color, black becoming white and white becoming black, hence the term *pattern reversal*. The response is the *pattern-reversal VEP*, or *PR-VEP*.

The checkerboard pattern is adjustable by the software of the computer, so that check size, brightness, contrast, and reversal rates can be adjusted. In routine clinical practice, these settings are seldom changed. Table 14.1 summarizes the stimulus and recording parameters for VEPs.

The response is changed by several stimulus-dependent factors:

- Size of the checks
- Size of the stimulated visual field
- Frequency of pattern-reversal
- Luminance of the display
- Contrast between bright and dark squares
- Fixation of the patient

FIGURE 14.1 *Checkerboard pattern produced by the video display for pattern-reversal visual-evoked potentials. The black-and-white squares reverse color.*

TABLE 14.1 Visual-evoked potential stimulus and recording parameters

Parameter	Value
Reversal rate	2/sec
Contrast	>50%
Check size	28′–32′ of arc
Field size	8° of arc
Number of trials	100–200
Recommended electrodes	
Midline occipital (MO)	5 cm above inion
Right occipital (RO)	5 cm right of MO
Left occipital (LO)	5 cm left of MO
Midline frontal (MF)	12 cm above nasion
Ear (A1)	Left ear or mastoid
Recommended montages	
Channel 1	LO-MF
Channel 2	MO-MF
Channel 3	RO-MF
Channel 4	MF-A1
Recording parameters	
Low-frequency filter	1.0 Hz (−3 dB)
High-frequency filter	100 Hz (−3 dB)
Analysis time	250 ms
Measurements	N75 latency from each eye
	P100 latency from each eye
	Interocular latency difference
	Amplitude (baseline to P100 peak or N75 to P100 peaks)
	Interocular amplitude ratio (larger/smaller)

Check size affects the latency and amplitude of the VEP. Size is measured in minutes of visual field arc where there are 60 minutes (60′) per degree of arc. The maximal response is elicited by a check size between 15′ and 60′. The wide range of check size is made possible by differences in other variables, notably size of the stimulus field. With smaller checks, the latency of the response is increased and the amplitude is decreased. The fovea is stimulated better by smaller checks, and the peripheral is stimulated better by larger checks. The recommended check size is 28′ to 32′ of visual field arc, which is a compromise between the two extremes.

Stimulus field size should be at least 8 degrees (8°) of the visual field arc, since approximately 80% of the response is generated by the central 8° of vision. A smaller field size has been recommended to increase sensitivity to subtle defects, but false positive rate is unacceptable. Visual acuity is the limiting factor in the presence of reduced stimulus field and reduced check size.

Reversal rate should be 2/sec, that is, an interstimulus interval of 500 ms. Faster reversal rates cause an increase in the latency of the major wave, P100. Rates faster than 5–7/sec produce the entrained driving response seen on routine EEG during photic stimulation.

Luminance is standardized, since low levels increase P100 latency and decrease amplitude. Standard computer monitors produce sufficient luminance for routine VEP testing, but this should be checked periodically. Note that the increasingly popular liquid-crystal diplay monitors often have a lag in luminance after being switched on, and some have slow refresh rates, compared to cathode-ray tube monitors. Pupillary diameter can affect effective retinal illumination, so marked interside differences in pupillary diameter must be considered in interpretation of the VEP. Although there are no specific recommendations for luminance levels, it is recommended that luminance level remain constant between studies.

Contrast between the light and dark squares must be greater than 50%. In practice, the contrast is much greater than this. Low contrast results in a delayed and lower amplitude P100.

Fixation on the target has more of an effect on amplitude than latency. Poor fixation results in reduction in amplitude, but as the response is detectable, there is much less effect on latency. Some individuals are able to voluntarily reduce the P100 amplitude sufficiently to make the P100 unidentifiable, partly through fixation.

Half-Field Pattern Reversal

Half-field stimulation is delivered to one eye at a time and one half-field, either right or left. The pattern-reversal technique is the same as that for full-field stimulation, with the checkerboard on one side and the screen blocked out. Comparing the responses with stimulation of the two half-fields tests the visual pathways behind the optic chiasm.

Magnetic resonance imaging and computed tomography provide excellent visualization of the retrochiasmal visual pathways and are superior to VEP for detection of lesions in these regions. Therefore, half-field stimulation is not commonly used for evaluation of patients with suspected brain pathology.

The left and right half-fields are alternately stimulated with the stimulated area at least one check-width from the fixation point. One eye is stimulated at a time. The montage is shown in Table 14.2.

Recording

Electrode Placement and Montage

Recommended electrode placements and montage are presented in Table 14.1. These are in keeping with the *Guidelines* (American Electroencephalographic Society, 1994), and are used by most neurophysiology labs.

Response from the midline-occipital (MO) electrodes is used for most VEP interpretation. The MO electrode is active and the midline-frontal (MF) electrode is the reference. Electrodes on either side of MO are right occipital (RO) and left occipital (LO); these can be helpful for interpretation of abnormal studies, but these responses are not used if MO responses are normal. Some laboratories use a Cz-Oz derivation, which is satisfactory, but labs should conform to the newer guidelines.

The LO and RO electrodes help with waveform identification, especially for half-field stimulation. For routine pattern-reversal stimulation, placement of electrodes is dictated by findings. If the waveforms are poor or unusual,

TABLE 14.2 Half-field visual-evoked potential stimulus and recording parameters

Parameter	Value
Electrodes	
Left occipital (LO)	5 cm left of midline occipital
Right occipital (RO)	5 cm right of midline occipital
Midline frontal (MF)	12 cm above the nasion
Left posterior temporal (LT)	10 cm left of midline occipital
Right posterior temporal (RT)	10 cm right of midline occipital
Montage for left half-field stimulation	
Channel 1	LO-MF
Channel 2	MO-MF
Channel 3	RO-MF
Channel 4	RT-MF
Montage for right half-field stimulation	
Channel 1	LT-MF
Channel 2	LO-MF
Channel 3	MO-MF
Channel 4	RO-MF
Measurements	P100 latency
	P100 amplitude
Calculations	Left half-field interocular latency difference
	Right half-field interocular latency difference
	Left eye half-field latency difference
	Right eye half-field latency difference
	Left half-field interocular amplitude ratio
	Right half-field interocular amplitude ratio
	Left eye half-field amplitude ratio
	Right eye half-field amplitude ratio

use of lateral electrodes, use of more anterior electrodes, or both is recommended to look for an unusual potential distribution.

Recording Parameters

Recording parameters are presented in Table 14.2. Filter settings are standard with a bandwidth of 1–100 Hz. Analysis time is 250 ms for most machines. Default recording parameters are set for EP machines, although they are changeable by the software.

Interpretation

Waveform Identification

Figure 14.2 shows normal VEPs. Inspection of the normal VEP reveals three identifiable waveforms: N75, P100, and N145. The P100 is a positive potential at about 100 ms, and is the only one used for VEP interpretation. The negative potentials at about 75 ms and 145 ms help with identification of the P100, but are too variable and inconsistent for routine interpretation.

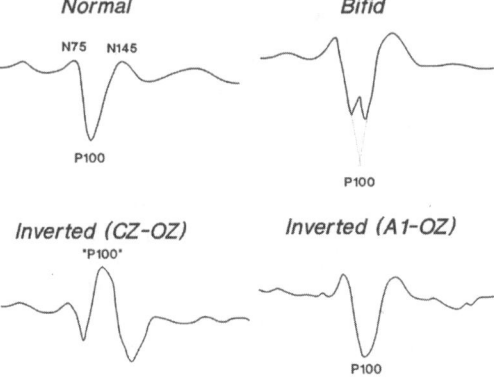

FIGURE 14.2 *Normal VEP potentials. Top left is a typical pattern with all three components labeled. Only the P100 is used for routine interpretation. The other three are normal variants.*

Variant Waveforms

The two most common variant waveforms are the *bifid* pattern and *inverted* waveform. Both are due to variation in anatomical orientation of the visual cortex and optic radiations.

Bifid Pattern
The P100 is split into two humps, which can be confusing. The neurophysiologist should treat this like a camel with one or two humps. Measure the hump from the peak of a single-hump beast, or interpolate between the humps for a two-hump beast. This interpolation is shown in Figure 14.2. If the interpolated P100 is normal, then the study is interpreted as normal, though comment on the variant waveform should be made in the impression.

Occasionally, patients will have a widely split P100 that is not amenable to extrapolation. This may be due to defects in the projection to the upper and lower segments of the calcarine cortex, possibly from visual field defects. Stimulation of only the lower half of the visual field may improve the bifid waveform, but the latency is often abnormal, anyway. There is controversy as to whether or not a bifid waveform should be considered abnormal on its own. The most conservative approach is to simplify the waveform by lower-half stimulation and localize the P100 by recording from Pz and Cz in addition to midline occipital derivations. Using these techniques, the rest is abnormal if the latency of the resultant waveform is increased.

Inverted Waveform
The inverted waveform is an artifact of a montage commonly used in recording the VEP. In some patients, the positivity of the VEP is shifted superiorally and anteriorally. Using a Cz-Oz montage, the Cz electrode is more positive than the Oz electrode, causing a reversed direction of the deflection.

N75 and N145 are reversed as well, and one of these could be misinterpreted as the P100. If the waveform is not typical of the normal VEP, recordings should be made from channels other than Cz-Oz, such as A1-Oz, A1-Pz, and A1-Cz. These will allow mapping of the topography of the VEP and aid greatly in identification of the P100.

Abnormalities

The types of abnormalities include:

- Unilateral prolongation of VEP latency
- Hemifield prolongation of VEP latency
- Absent VEP
- Reduced amplitude of VEP

Unilateral prolongation of VEP suggests slowing of conduction in one optic nerve. Hemifield prolongation of VEP latency suggests a defect in the conduction behind the optic chiasm. Table 14.3 shows normal data for VEP.

Clinical Correlations

Optic Neuritis

Optic neuritis typically increases the latency of the P100 of the PR-VEP. If the optic neuritis is purely unilateral, then the increase is usually also unilateral. A prolonged latency of the P100 from an asymptomatic eye suggests a previous subclinical episode of optic neuritis. A sample VEP from a patient with optic neuritis is shown in Figure 14.3.

Multiple Sclerosis

Approximately 15% of patients with optic neuritis will later develop other findings leading to the diagnosis of multiple sclerosis. In patients with optic neuritis, SEP is frequently performed to look for clinically silent lesions of the spinal cord. VEP testing may be done to document the visual defect in the clinically affected eye and to evaluate the optic nerve of the unaffected eye. Abnormal VEP latency is present in about 40% of patients with multiple sclerosis who do not have a history of optic neuritis. Virtually all patients

TABLE 14.3 Visual-evoked potential normal data

Measurement	Values
Pattern-reversal visual-evoked potential	
Latency	117 ms
Interocular latency difference	6 ms
Amplitude	3 µV
Interocular amplitude difference	5.5 µV
Flash visual-evoked potential	
Latency	132 ms
Interocular latency difference	6 ms

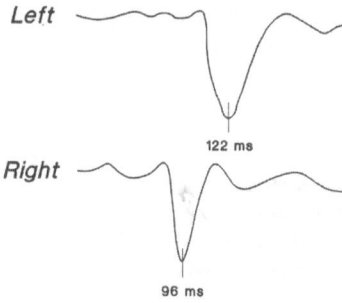

FIGURE 14.3   *VEP in optic neuritis. Left optic neuritis produces a delay in the VEP produced by stimulation of that eye.*

with a history of optic neuritis have either an absolute increase in the latency of the P100 from the affected eye or an abnormal interside difference in latency, if the absolute latencies are normal.

Tumors

Tumors arising in the region of the anterior optic pathways produce compression of optic neuritis and/or chiasm. This results in visual field defects that affect each eye differently. The VEP is almost always abnormal, but the correlation between visual acuity and degree of VEP abnormality is poor. Typical abnormalities are alterations in absolute or interside latencies of the P100 and changes in wave morphology and amplitude. Latency changes are more reliable than morphology or amplitude changes.

Tumors affecting the posterior visual pathways are less likely to affect the VEP. Full-field pattern-reversal VEP testing is usually normal in patients with dense hemianopia. The use of electrodes lateral to MO may reveal an amplitude asymmetry with the higher amplitude ipsilateral to the side of the lesion, but amplitude asymmetries may be present in normal individuals. Half-field stimulation reveals abnormalities in some individuals; however, the sensitivity and specificity are not good enough to justify using this technique to screen for posterior lesions. Imaging techniques should be used.

Pseudotumor Cerebri

Patients with pseudotumor cerebri have increased intracranial pressure that is not due to a mass lesion, venous thrombosis, or other structural defect. If untreated, the increased pressure can produce visual loss. If treatment is effective, the visual loss can improve, however permanent deficits result if there has been long-standing increased pressure.

Most patients with pseudotumor cerebri have normal VEPs. A few patients are described with abnormalities in association with incipient visual loss; however, VEPs should not be used as screening procedures for increased intracranial pressure.

Functional Visual Loss

The PR-VEP is frequently used to evaluate patients with suspected functional visual loss. Few individuals can voluntarily suppress the VEP. An intact VEP usually suggests continuity of the visual pathways but does not fully exclude cortical blindness. Caution is needed in using PR-VEP for the diagnosis of functional visual loss.

A normal flash VEP indicates continuity of the visual pathways only to the lateral geniculate. An intact flash response is expected with lesions of optic radiations and visual cortex. Great care should be exercised in interpretation of latency abnormalities of the flash VEP.

Ocular Disorders

Many ocular disorders cause abnormalities in the PR-VEP. The effect of impaired visual acuity and effective retinal illumination on the VEP have been discussed previously. The VEP is not ordinarily used for the diagnosis of these disorders. Although some patients with glaucoma have an increased latency and reduced amplitude of the P100, a normal VEP cannot be interpreted as indicating normal intraocular pressure. The effects of ocular and retinal disorders on the VEP are of interest only in the interpretation of VEPs used to evaluate disorders at and behind the optic nerve.

Cortical Blindness

Some patients with documented cortical blindness have normal PR-VEPs. The use of smaller check size may help to bring out abnormalities, however, this is not routinely done in most neurodiagnostic laboratories.

15 Brainstem Auditory-Evoked Potentials

Brainstem auditory-evoked potentials (BAEPs) are used mainly for evaluation of auditory circuits in the brainstem. In modern practice, neonates are predominantly studied, although adults are also examined for evaluation of reduced hearing and suspected disorders of the brainstem. Traditionally, BAEP has been a particularly cost-effective screening test for acoustic neuroma. Intraoperative monitoring of BAEPs is also occasionally performed during surgery on the posterior fossa. The BAEP is less helpful for evaluation of multiple sclerosis (MS) than visual-evoked potentials (VEP) and somatosensory-evoked potentials (SEP).

Methods

Stimulus

Earphones
Headphones are placed on the ears for delivery of the auditory stimulus. For adults and older children, large headphones that surround the external ear are used. These provide excellent sound delivery and block out much ambient noise. Young children and infants cannot use these headphones because they may collapse the external auditory canal (EAC), so small earphones are inserted into the EAC. Both types of phone should be provided with the EP machine.

Types of Stimulus
The EP apparatus delivers signals to the phone that produce one of a number of different sounds. The most commonly used sounds are:

- Clicks
- Tones
- White noise

Clicks are routinely used as a stimulus for routine BAEP. The electrical signal is a square wave with a rapid upstroke, plateau, and rapid downstroke to neutral voltage. Since the direction of voltage change is opposite to the rising and falling phases of the voltage signal, the earphone diaphragm moves toward the eardrum with one phase and away from the eardrum with the other. When the diaphragm moves toward the eardrum, the air in the canal is compressed or condensed; therefore, this is termed condensation. When the diaphragm moves away from the eardrum, the air in the canal is decompressed, or rarefied. Therefore, this is termed rarefaction. The distinction is more than semantic; the response can differ substantially between condensation and rarefaction clicks, so the response is often averaged separately for the two modes. Of course, there has to be a condensation movement of the diaphragm for every rarefaction, so both are delivered. Interpretation of the BAEPs to rarefaction and condensation clicks is discussed below.

Tones are produced by a sine wave, but are seldom used in routine BAEP. More commonly, pure tones are used for audiometry, where hearing thresholds are determined for different frequencies. Some disorders can produce a predominantly high-frequency hearing loss, as with ototoxic drugs. This would be less well detected by conventional BAEP than by pure-tone audiometry.

White noise is not used as a stimulus, but is delivered to the nonstimulated ear as a masking sound. This masking sound reduces spread of conduction of the stimulus sound delivered to the opposite ear, so that the stimulation is not bilateral. Without this, air or bone conduction of the stimulus can produce bilateral stimulation. The sound is similar to that heard by a radio that is tuned to a frequency without a broadcasting station. Almost every frequency is represented in the signal. The reason it is called white noise is because the broad spectrum of white noise is analogous to white light, where the white appearance is due to the presence of all colors (frequencies or wavelengths) of light.

Stimulus Rate
Clicks are delivered at a rate of 8–10/sec. This allows for reproducible identification of all waves. Waves I, II, VI, and VII are reduced in amplitude at faster frequencies.

Stimulus Intensity
Terminology for expression of stimulus intensity is confusing. The *Guidelines* (American Electroencephalographic Society, 1994) recommend that intensity be expressed in units of decibels peak equivalent sound pressure level, or dB pe SPL. This is measured directly using a sound meter with a constant stimulus of the same frequency and amplitude as the test stimulus to be measured. The dB scale is a log scale, so 0 dB is defined as a pressure of 20 micropascals. An alternative scale occasionally used is in reference to normal hearing threshold, or dB HL (hearing level). Zero dB HL is the threshold for hearing for a population of normal people. This is approximately equivalent to 30 dB pe SPL. Sensation level, or dB SL, is in reference to the ear being tested; 0 dB SL is the threshold for that ear.

The *Guidelines* recommend stimulus intensities between 40 and 120 dB pe SPL. Many EP machines do not give stimuli louder than that recommended for routine BAEP testing. Intensity is set at 65 dB SL or HL. Reducing stimulus intensity is necessary only if waveform identification is difficult. With decreasing stimulus intensity, waves II and VI are reduced more than the other waves, allowing for more accurate identification of waves I, III, and V.

Recording

Electrodes are placed in the following positions:

- A1—behind the left ear
- A2—behind the right ear
- Cz—vertex

These electrode placements are identical to those for routine EEG recording. Recommended montages for recording of BAEP are:

- Channel 1 = Cz-Ai
- Channel 2 = Cz-Ac

where Ai is the ipsilateral ear and Ac is the contralateral ear. Therefore, for each ear being stimulated, the first channel is from the ipsilateral ear in reference to the vertex and the second channel is from the contralateral ear in reference to the vertex.

Recording parameters are summarized in Table 15.1.

Interpretation

Figure 15.1 shows a normal BAEP. The appearance can differ slightly, but the waves of interest are all convex up and have the same multilobed appearance.

Waveform Identification

The waves routinely analyzed in BAEP testing are numbered I through V. Waves VI and VII are also identified but not used in interpretation. Waves I and V should be identified first. Wave I is generated by the distal portion of the acoustic nerve and is approximately 2 ms after the stimulus. Wave I identification is aided by recording from a contralateral electrode derivation, it is the only wave present on ipsilateral but not contralateral recording.

Wave V may be generated by projections from the pons to the midbrain. There are several criteria for identifying wave V. It normally appears at approximately 6 ms and is often combined with wave IV into a single complex waveform. Wave V is also the first waveform whose falling edge dips below the baseline.

TABLE 15.1 Brainstem auditory-evoked potential stimulus and recording parameters

Parameter	Value
Stimulus parameters	
Rate	8–10/sec
Intensity	115–120 dB pe SPL
Duration of pulse	100 μs
Stimulus character	Monaural
	Contralateral masking noise (60 dB pe SPL)
Stimulus polarity	Rarefaction or condensation, summed independently
Number of trials	1,000–4,000
Recording parameters	
Recommended montages	
Channel 1	Cz-Ai
Channel 2	Cz-Ac
Low-frequency filter	10–30 Hz (−3 dB)
High-frequency filter	2,500–3,000 Hz (−3 dB)
Analysis time	10–15 ms
Measurements	Wave I peak latency
	Wave III peak latency
	Wave V peak latency
	Wave I amplitude
	Wave V amplitude
Calculations	I–III interpeak interval
	III–V interpeak interval
	I–V interpeak interval
	Wave V/I amplitude ratio

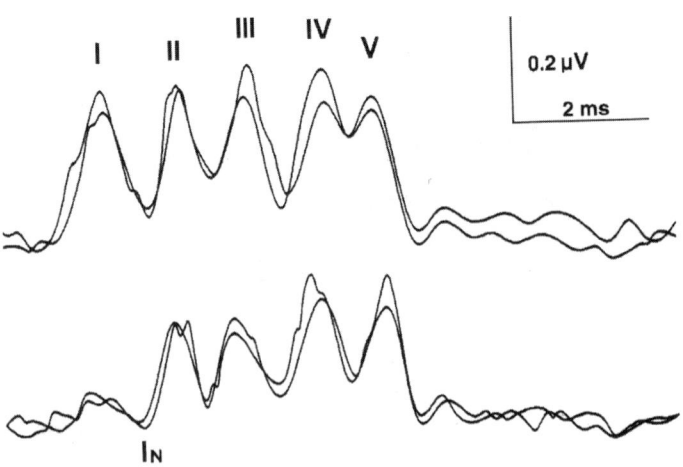

FIGURE 15.1 Normal BAEP in response to stimulation of the right ear. Response from the right is on top, response from the left is on the bottom. Note the absence of wave I in the contralateral recording.

The wave III–V complex has a wider separation with recording from the ipsilateral ear. This means that the contralateral wave IV is of slightly shorter latency and wave V is of slightly longer latency. Wave V is the last to disappear as stimulus intensity is decreased.

Wave III is thought to be generated by the projections from the superior olive through the lateral lemniscus. It is the major peak between waves I and V.

Data Analysis

Latency is a more important measure than amplitude in the interpretation of BAEP data. The most important measurements are:

- Wave I latency
- Wave III latency
- Wave V latency

From these data, the following are calculated:

- I–III interpeak interval
- III–V interpeak latency
- I–V interpeak latency

Table 15.2 presents normal BAEP data. Interpretation of the data is as follows:

- *Increased wave I latency*: Damage to the most distal portion of the acoustic nerve. Acoustic neuromas seldom affect wave I.
- *Increased I–III interpeak interval*: Defect in the pathway from the proximal eighth nerve into the inferior pons. The lesion may be either in the nerve or in the brainstem. This is the most common abnormality found in patients with acoustic neuromas.
- *Increased III–V interpeak interval*: Defect in the conduction between the caudal pons and the midbrain.
- *Increased I–III and wave III–V interpeak intervals*: Lesion affects the brainstem at and above the caudal pons with or without involvement of the acoustic nerve. In most instances, this is due to a prominent lesion in the pons.
- *Absence of wave I with normal III and V*: May indicate a peripheral hearing disorder, with the caveat that conduction in the caudal pons cannot be excluded.

TABLE 15.2 Brainstem auditory-evoked potential normal data

Waveform	Male	Female
Wave I latency	2.10	2.10
I–III interpeak interval	2.55	2.40
III–V interpeak interval	2.35	2.20
I–V interpeak interval	4.60	4.45
I–V interside difference	0.05	0.05
Wave V/I amplitude ratio	0.05	0.05

- *Absence of wave III with normal waves I and V*: Normal, but if the wave I–V interval is prolonged, then a lesion affecting conduction somewhere from the eighth nerve to the midbrain is suspected.
- *Absence of wave V with normal waves I and III*: This is uncommon, but when present indicates a lesion affecting the auditory pathways above the caudal pons. This is considered an extreme prolongation of the wave III–V interval.

Abnormalities

Table 15.3 presents guides to interpretation of abnormalities on BAEP.

Lesions of the lower brainstem or acoustic nerve can produce increased I–III interpeak interval (Figure 15.2). This could be due to acoustic neuroma but is more likely to be due to lesion at the cerebellopontine angle.

Lesion of the upper brainstem can produce increased III–V interpeak interval (Figure 15.3). This can be due to stroke, mass lesions, or demyelinating disease. However, BAEP is remarkably insensitive for diagnosis of MS.

Pediatric BAEPs

The most common use of BAEPs in pediatric patients is for assessment of brainstem function in premature infants. Assessment of hearing is also performed, and for this, audiometry is chiefly used.

TABLE 15.3 Interpretation of brainstem auditory-evoked potential findings

Finding	Interpretation
Increased wave I latency	Lesion of distal portion of acoustic nerve
Increased I–III interpeak interval	Lesion of pathway from proximal cranial VIII to pons, either in the nerve or in the brainstem (e.g., acoustic neuroma)
Increased III–V interpeak interval	Lesion between caudal pons and midbrain
Increased I–III and III–V interpeak intervals	Lesion affecting brainstem above caudal pons plus either the caudal pons or acoustic nerve
Absence of wave I with normal III and V	Peripheral hearing disorder Conduction in the caudal pons cannot be evaluated
Absence of wave III with normal I and V	Normal study
Absence of wave V with normal I and III	Lesion above the caudal pons, considered an extreme of wave III–V interpeak interval prolongation
Absence of all waves	Severe hearing loss

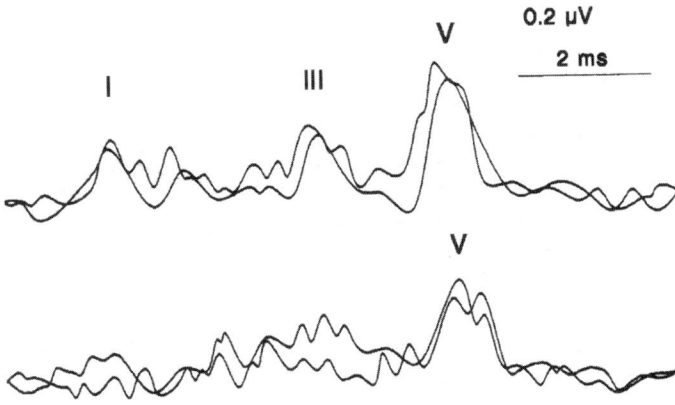

FIGURE 15.2 *Increased I–III interpeak interval of BAEP. Patient with a lesion of the lower brainstem has produced an increased I–III interpeak interval, and failure to identify a definite wave III on the contralateral side.*

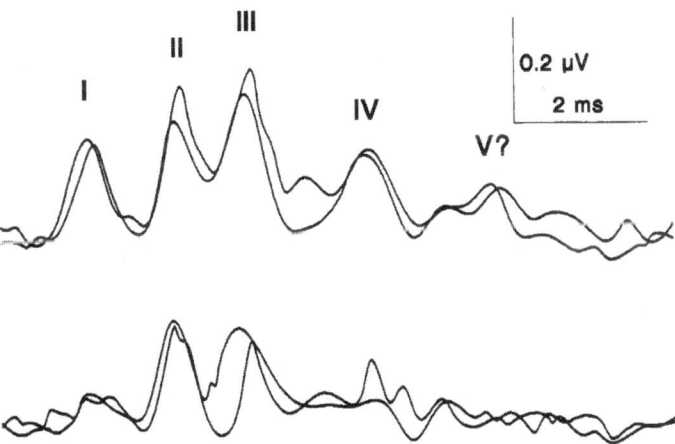

FIGURE 15.3 *Increased III–V interpeak interval of BAEP. Patient with a lesion of the upper brainstem producing an increased III–V interpeak interval and interfering with recording of wave V from the contralateral side.*

Childhood

BAEP in children is mainly used to assess hearing in patients who cannot cooperate with conventional hearing tests. An abnormal BAEP is usually associated with abnormalities on behavioral testing of hearing; however, a normal BAEP does not guarantee normal hearing. If the lesion is of the peripheral auditory structures, threshold may be increased, but there may not be a change in the wave I–V interpeak interval.

Abnormal BAEPs are seen in several disorders of childhood, including phenylketonuria, maple syrup urine disease, nonketotic hyperglycinemia, and Leigh's disease. Wave I–V interpeak interval is typically increased. BAEP testing is not important in the diagnosis of these disorders, however, because BAEP abnormalities are not disease specific.

Neonates

Methods are slightly different in neonates than in older children. Earphones are used for neonates rather than headphones, because the latter may collapse the external auditory canal. Sedation may be required, and is best accomplished by chloral hydrate, although meperidine plus secobarbital or meperidine plus promethazine are also used. Sedation does not affect the short-latency EPs such as BAEP.

BAEP is performed on infants to evaluate respiratory and feeding dysfunction, particularly with suspected perinatal asphyxia and in premature infants. The wave I–V interpeak interval is prolonged in premature infants and may be related to delayed maturity of brainstem nuclei and pathways. There is an almost linear relationship between the decline in wave I–V interpeak latency and the reduction in apnea frequency.

Wave I–V interpeak interval is increased in term newborns who have experienced episodes of total asphyxia with subsequent damage to brainstem nuclei. The mortality and neurologic morbidity in such newborns is high. Newborns who have experienced prolonged partial asphyxia sustain mainly hemispheric damage, and the wave I–V interval may be normal despite a poor neurologic outcome.

Clinical Correlations

Acoustic Neuroma

The most sensitive finding for the diagnosis of acoustic neuroma is prolongation of wave I–III interpeak interval. If there is difficulty in obtaining wave I, the technician should place an electrode in the external auditory canal for better recording. Alternatively, electrocochleography can aid with identification of wave I. In patients with very large tumors, there may be such severe damage that there are no reproducible waves after wave I. In early cases, the BAEP has been abnormal, when imaging revealed an acoustic neuroma. A sample finding with acoustic neuroma is shown in Figure 15.2.

Brainstem Tumor

BAEP test results are abnormal in most patients with intrinsic tumors of the brainstem. This is especially true in patients with pontine involvement. The usual abnormality is delay or loss of waves III and V and increased wave I–III and wave III–V interpeak interval.

Stroke

BAEP test results are abnormal in most patients with brainstem stroke. A few patients with extensive brainstem infarctions have been reported to have normal BAEPs, however. In some of these patients the amplitudes of the waveforms were low; however, amplitude abnormalities are not emphasized in the interpretation of BAEPs.

Approximately 50% of patients with transient ischemic attacks affecting the posterior circulation have latency abnormalities, and 50% of patients who recover from definite brainstem strokes have normal BAEPs. A sample recording from a patient with a brainstem lesion is shown in Figure 15.3.

Multiple Sclerosis

BAEP testing is less sensitive than VEPs and SEPs for detection of clinically unsuspected lesions in patients being evaluated for MS. The usual abnormalities are reduction in wave V amplitude and increased wave III–V interpeak interval.

Most abnormalities are asymmetric, affecting the response from only one ear. Caution in the interpretation of amplitude abnormalities is recommended. BAEP testing cannot distinguish a demyelinating disease from tumors or infarction.

Coma and Brain Death

The Medical Consultants on the Diagnosis of Death (1981) cited BAEP testing as a confirmatory test, along with EEG and radionucleotide brain scan. The commission criteria are enumerated in Chapter 7.

BAEP test results are consistent with brain death if there are no reproducible waves after wave I. Wave II may be intact in less than 10% of brain dead patients, reinforcing the hypothesis that wave II is generated by the intracranial portion of the eighth nerve. The presence of wave II is consistent with brain death in a patient who otherwise fulfills all other clinical criteria and has no subsequent waves on the BAEP potential.

Audiometry

Audiometry uses BAEP to assess the function of the middle and inner ear rather than the brainstem. The technique is similar to that used for conventional BAEP except that special attention is given to the wave V latency as a function of stimulus intensity. With increasing intensity, the wave V latency becomes progressively shorter. Stimuli are delivered at stimulus intensities of 20, 40, 60, and 80 dB greater than threshold. Because of the nature of the decibel scale, this is essentially a semilog plot (Figure 15.4).

The relationship between stimulus intensity and wave V latency (latency–intensity curve) is linear in most individuals, with higher intensities producing shorter latencies. *Conductive hearing loss* does not change the slope of this relationship but prolongs the latency at each intensity. Therefore, the curve is shifted upward. The response looks as if the intensities were turned down at every point, which is essentially what occurs with conductive hearing loss. Sensorineural hearing loss produces a curve with two slopes. At low intensities, there is decreased responsiveness of the end-organ, so that for a given intensity the wave V latency is prolonged. With increases in intensity, there is more recruitment of nerves than normal, so that the slope of the

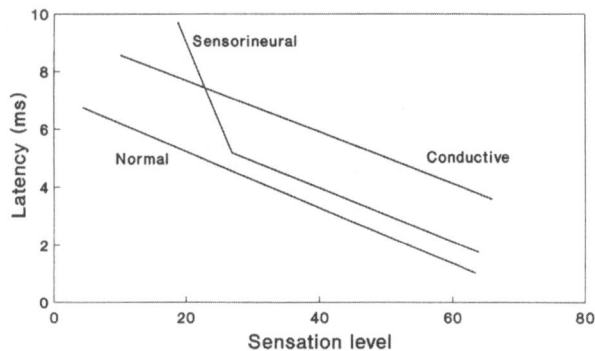

FIGURE 15.4 *Audiometry. Wave V latency of the BAEP as a function of stimulus intensity. Normal patients have an inverse relationship. Patients with conductive hearing loss have a response curve that is parallel to the normal curve, but pushed up and to the right---lower sensitivity. Patients with sensorineural hearing loss have a two-sloped curve.*

curve is steeper. At high intensities, sufficient recruitment has occurred such that the latency may be normal. At this point, the slope reverts to normal. This L-shaped curve is typical for sensorineural hearing loss.

Audiometry is useful for evaluating patients for hearing loss when the localization of the lesion is in doubt. Audiometry can also be used to follow patients receiving chemotherapeutic agents that cause ototoxicity.

16 Somatosensory-Evoked Potentials

The somatosensory-evoked potential (SEP) is the response to electrical stimulation of peripheral nerves. Stimulation of almost any nerve is possible, although the most commonly studied nerves are:

- Median
- Ulnar
- Peroneal
- Tibial

Brief electric pulses are delivered to the peripheral nerve with the cathode proximal to the anode. The stimulus cannot selectively activate sensory nerves, so a small muscle twitch is seen. There are no effects of the retrograde motor volley in the motor nerves on central projections of the sensory fibers.

Intensity of the stimulus is adjusted to activate low-threshold myelinated nerve fibers, which in the motor fibers elicits a small twitch of the innervated muscles. The compound action potential is conducted through the dorsal roots and into the dorsal columns. The impulses ascend in the dorsal columns to the gracile and cuneate nuclei where the primary afferent fibers synapse on the second-order neurons. These axons ascend through the brainstem to the thalamus. Thalamocortical projections are extensive, as are secondary intracortical associative projections. Table 16.1 shows common stimulus and recording parameters.

The recording is made from various levels of the nervous system, including:

- Afferent nerve volley
- Spinal cord
- Brain

Recording of the afferent nerve volley ensures that the stimulus is adequate, and determines whether there is a defect in peripheral conduction that would interfere with interpretation of the central waveforms. Recordings from the spinal cord measure electrical activity in white-matter tracts and relay nuclei. Recordings from brain will measure the projections from relay nuclei to the cortex. Short-latency responses are used for clinical interpretation; long-latency responses are too variable to be helpful.

TABLE 16.1 Somatosensory-evoked potential common stimulus and recording parameters

Parameter	Value
Stimulus rate	45–7/sec
Low-frequency filter	5–30 Hz (−3 dB)
High-frequency filter	2,500–4,000 Hz (−3 dB)

The anatomy of the projections from the peripheral nerve to cortex is complex. Unfortunately, there is not a precise anatomical correlate to each of the waves in the SEP. This complicates interpretation, unfortunately.

SEPs are particularly useful for evaluation of function of the spinal cord. Lesions of the cord may be invisible to routine imaging, including magnetic resonance imaging and myelography, yet may have devastating effects on cord function. Transverse myelitis, multiple sclerosis (MS), and cord infarction are only three of the potential causes that can be missed on structural studies.

Although almost any nerve can be studied, median and tibial will be discussed here. Additional information on other techniques can be seen in *Spehlmann's Evoked Potential Primer* (Misulis and Fakhoury, Butterworth-Heinemann, 2001).

Median SEP

Methods

Stimulating electrodes are placed over the median nerve at the wrist with the cathode proximal to the anode. Stimuli are square-wave pulses given at rates of 4–7/sec.

Recording electrodes are placed at the following locations:

- Erb's point on each side (EP)
- Over the second or fifth cervical spine process (C2S or C5S)
- Scalp over the contralateral cortex (CPc) and ipsilateral cortex (CPi)
- Noncephalic electrode (Ref)

Erb's point is 2–3 cm above the clavicle, just lateral to the attachment of the sternocleidomastoid muscle. Stimulation at Erb's point produces abduction of the arm and flexion of the elbow. The second and fifth spinous processes are identified by counting up from the seventh, notable by its prominence at the base of the neck. CPc and CPi are scalp electrodes halfway between C3 and P3 or C4 and P4, where CPc is contralateral to the stimulus and CPi is ipsilateral to the stimulus. These electrodes are over the motor-sensory cortex. EPi is Erb's point ipsilateral to the stimulus. The recommended montage is:

- Channel 1: CPc-CPi
- Channel 2: CPi-Ref

- Channel 3: C5S-Ref
- Channel 4: Epi-Ref

Details of the recommended stimulus and recording parameters are presented in Table 16.2, including analysis time and filter settings. These parameters are usually not changed for individual studies. The number of trials averaged for adequate waveform identification is 500–2,000, but more may occasionally be needed. Duplicate trials overlaid on the display help greatly with waveform identification.

Waveform Identification and Interpretation

Figure 16.1 shows a normal median SEP. The recorded potentials are from Erb's point, spinal cord, and scalp. The waves of interest are:

- N9 = from the Epi channel
- P14 = from the neck channel
- N20 = from scalp channels

N9: The potential recorded from Erb's point is sometimes called EP, but because of the obvious abbreviation confusion, N9 is a preferable designation. The N9 potential is a compound action potential from the axons stimulated by the median nerve stimulation. While the N9 potential includes both orthograde sensory nerve potentials and retrograde motor nerve potentials, all of the subsequent waves are due to sensory activation, alone.

TABLE 16.2 Median nerve somatosensory-evoked potential stimulus and recording parameters

Parameter	Value
Number of trials	500–2,000
Analysis time	40 ms
Bandpass	30–3,000 Hz (−6 dB)
Stimulating electrodes	
Cathode	Median nerve 2 cm proximal to the wrist crease
Anode	2 cm distal to cathode
Recording electrodes	
CPc	Contralateral scalp, between C3/4 and P3/4
CPi	Ipsilateral scalp, between C3/ and P3/4
C5S	Over C5 spinous process
Epi	Erb's point ipsilateral to the stimulus
Ref	Noncephalic reference, such as distal arm
Montage	
Channel 1	CPc-CPi
Channel 2	CPi-Ref
Channel 3	C5S-Ref
Channel 4	EPi-Ref
Measurements	Peak latency of N9 potential
	Peak latency of P14 in neck–scalp channel
	Peak latency of N20 in scalp–scalp or scalp–ear channel

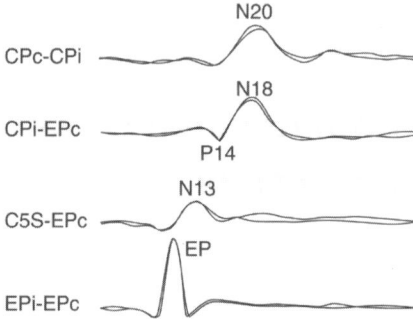

FIGURE 16.1 *Normal median nerve SEP.*

P14: The neck potentials include N13 and P14, with the latter used for clinical interpretation. The origin is thought to be the caudal medial lemniscus.

N20: Scalp potentials include N18 and N20, but the latter is used for clinical interpretation. Origin is thought to be the thalamocortical radiations.

Calculated Data Table 16.3 shows normal SEP data and Table 16.4 shows probable waveform origins for the SEP potentials. When the N9, P14, and N20 measurements have been made, calculations are made to assess conduction between the regions. N9–P14 interval represents the time for conduction between the brachial plexus and cervical spine. P14–N20 interval represents the conduction between the cervical spine and the brain. This is called the *brain conduction time* (BCT).

The median SEP is used for several indications including transverse myelitis, MS, thoracic outlet syndrome, and intraoperative monitoring during

TABLE 16.3 Somatosensory-evoked potential normal data

Wave	Latency	Interside difference
Median nerve		
N9/EP	11.80	0.87
P14	16.10	0.70
N20	21.50	1.20
P14–N20	7.10	1.20
Peroneal nerve		
LP	13.50	0.50
P27	31.80	2.24
N33	40.70	5.96
LP–P27 interval	19.38	2.30
Tibial nerve		
LP	2.10	1.20
P37	41.70	1.40
LP–P37 interval	20.50	1.50

TABLE 16.4 Somatosensory-evoked potential waveform origins

Wave	Origin
Median	
N9/EP	Afferent volley in plexus
N13	Dorsal horn neurons
P14	Caudal medial lemniscus
N18	Brainstem and thalamus?
N20	Thalamocortical radiations
Tibial	
LP	Dorsal roots and entry zone
N34	Brainstem and thalamus?
P37	Primary sensory cortex

carotid endarterectomy. Interpretation of abnormalities is as follows:

- *Delayed N9 with normal NP–P14 and P14–N20 intervals*: Lesion in the somatosensory nerves at or distal to the brachial plexus.
- *Increased N9–P14 interval with normal P14–N20 interval*: Lesion between Erb's point and the lower medulla.
- *Increased P14–N20 with normal N9–P14 interval*: Lesion between the lower medulla and the cerebral cortex.

Amplitude abnormalities should be interpreted with caution. A marked asymmetry can be caused by a lesion affecting some but not all of the afferent fibers. If the lesion is not sufficiently severe to produce a latency change; however, the study is probably normal. An absent N20 is abnormal when N9 and P14 are present. If P14 is absent but N9 and N20 are normal and the N9–N20 interval is normal, the study indicates a lesion between the brachial plexus and medulla, but no statement can be made of brain conduction.

Tibial SEP

Methods

Stimulus settings are similar to those used for median SEP, and details of the stimulus and recording parameters are presented in Table 16.5. The proximal stimulating electrode (cathode) is placed at the ankle between the medial malleolus and the Achilles tendon. The anode is placed 3 cm distal to the cathode. A ground is placed proximal to the stimulus electrodes, usually on the calf. Stimulus intensity is set so that each stimulus produces a small amount of plantar flexion of the toes.

Recording electrodes are placed as follows:

- CPi = Ipsilateral cortex between C3 and P3 or C4 and P4
- CPz = Midline between Cz and Pz
- FPz = Fpz position of the 10–20 electrode system
- C5S = Over the C5 spinous process

TABLE 16.5 Tibial nerve somatosensory-evoked potential stimulus and recording parameters

Parameter	Value
Number of trials	1,000–4,000
Stimulating electrodes	Behind medial malleolus
Analysis time	60 ms
Bandpass	30–3,000 Hz (–6 dB)
Recording electrodes	
CPi	Ipsilateral cortex between C3/4 and P3/4
CPz	Midline between Cz and Pz
Fpz	Fpz position of the 10–20 system
C5S	Over the C5 spinous process
T12S	Over the T12 spinous process
Ref	Noncephalic reference
Montage	
Channel 1	CPi-Fpz
Channel 2	CPz-Fpz
Channel 3	Fpz-C5S
Channel 4	T12S-Ref
Measurements	LP latency
	P37 latency
Calculation	LP–P37 interpeak interval

- T12S = Over the T12 spinous process
- Ref = Noncephalic reference

The montage used for routine tibial SEP is as follows:

- Channel 1: CPi-Fpz
- Channel 2: Cpz-Fpz
- Channel 3: Fpz-C5S
- Channel 4: T12S-Ref

Waveform Identification and Interpretation

Figure 16.2 shows a normal tibial SEP. The recorded waveforms with tibial stimulation are from the afferent fibers in the dorsal roots and dorsal root entry zone, spinal cord, and brain. The waves used for interpretation are:

- LP = from the T12S channel
- P37 = from the scalp channel

LP: The lumbar potential (LP) is thought to arise from the afferent nerve volley in the dorsal roots and dorsal root entry zone. Identification of the LP is usually easy. Patients with peripheral neuropathy may have desynchronization of the afferent volley, such that the amplitude of the nerve potentials may be low or inconsistent.

P37: P37 is a positive potential at about 37 ms that is seen from the scalp channels. The origin is thought to be the primary sensory cortex. N34

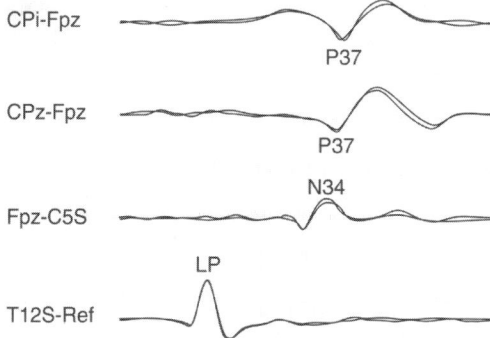

FIGURE 16.2 *Normal tibial nerve SEP.*

precedes P37, but is not used for clinical interpretation. Identification of the N34 and P37 is facilitated by overlapping averaged traces, which is currently recommended to aid visual interpretation of EPs in general. N34 is the main negative wave in the FPz-C5S derivation and is preceded by a small positive wave that is not used for interpretation (P31). The P37 is the major positive wave in the Cpi-Fpz and CPz-Fpz channels. The amplitude may be very different between these channels, reflecting an anatomic variance in waveform distribution.

Calculated Data   The LP–P37 interval is the time from the cauda equina to the brain. This is called the *central conduction time* (CCT).

Cervical Spine Data   Cervical potentials can occasionally be recorded during tibial SEP, but these potentials are too inconsistent and too variable to be used for clinical interpretation. Many laboratories do not even try to record these potentials, rather using median SEP for clinical use.

Interpretation of Abnormalities   Interpretation of tibial SEPs parallels interpretation of median SEPs. Absence of waveforms and increased interpeak intervals are the most important interpretive data. Abnormalities of tibial SEPs are interpreted as follows:

- *Prolonged LP with normal LP–P37 interval*: Peripheral or distal lesion. Peripheral neuropathy is most likely, but the slowing could be in the cauda equina.
- *Normal LP with prolonged LP–P37 interval*: Abnormal conduction between the cauda equina and the brain. Median SEP is required to localize the abnormality to the spinal cord. Normal median SEP indicates a lesion below the mid-cervical cord. Prolonged median SEP indicates a lesion above the mid-cervical cord. A second lesion below the cervical cord cannot be ruled out, however, since the P37 latency is already prolonged by the higher lesion.

TABLE 16.6 Somatosensory-evoked potential findings in selected disorders

Disorder	Finding
Arm	
AIDP (Guillain-Barré syndrome)	Proximal conduction (clavicular spinal) slowed more than peripheral conduction; normal BCT
Amyotrophic lateral sclerosis	Normal
Anterior spinal artery syndrome	Normal; posterior columns are spared
Brachial plexus lesions	Delayed cervical potential; BCT is normal
Brain death	Absent scalp potentials
Brainstem stroke	Normal in lateral medullary syndrome; in infarction involves medial lemnisci, absent or delayed scalp potential
Cervical cord lesion	Loss of cervical potential; clavicular potential normal; increased BCT
Cervical radiculopathy	Usually normal
Chronic renal failure	Delays of all peaks and reduced amplitude; BCT usually normal
Congenital insensitivity to pain	Normal
Creutzfeldt–Jakob disease	Variable results; may be normal or of increased or decreased amplitude
Friedreich's ataxia	Slowed peripheral conduction, may also have increased BCT
Head injury	Normal or abnormal; a normal response is of moderate prognostic value
Hemispherectomy	Abolish scalp potentials
Hepatic encephalopathy	Increased BCT
Hereditary pressure-sensitive palsy	Slowed peripheral conduction
Hyperthyroidism	Increased amplitude of scalp
Leukodystrophies	Reduce or abolish cervical potentials; delay or abolish scalp potentials
Minamata disease (mercury)	Absent scalp potential; cervical potential normal
Multiple sclerosis	Increased clavicular–spinal or BCT
Myoclonus epilepsy	Increased amplitude of scalp potential
Parietal lesion	Absent or low-amplitude scalp potential; less delay with cortical than subcortical lesion
Perinatal asphyxia	Absent, low-amplitude, or increased latency of the scalp potential; degree of abnormality correlates with extent of damage
Persistent vegetative state	Usually absent or delayed scalp potential
Polyneuropathy	Delayed peaks Normal BCT; clavicular–cervical time may be increased
Reye's syndrome	Abnormal early; return of peaks indicates a good prognosis
Subarachnoid hemorrhage	Low-amplitude or delayed scalp potential
Thoracic outlet syndrome	Delayed cervical potential; clavicular potential may be delayed or low amplitude
Tourette's syndrome	Normal
Leg	
Adrenoleukodystrophy	Increased CCT
Charcot–Marie–Tooth disease	Normal; peripheral slowing
Diabetes mellitus	Peripheral slowing; occasional central slowing
Friedreich's ataxia	Increased CCT
Multiple sclerosis	Increased CCT, normal peripheral conduction
Myotonic dystrophy	Slowed peripheral conduction
Radiculopathy	Usually normal
Spinal cord injury	Often abnormal, early return or normal response indicates a favorable prognosis
Subacute combined degeneration	Cortical potentials delayed or absent

AIDP = acute inflammatory demyelinating polyradiculoneuropathy; BCT = brain conduction time; CCT = central conduction time.

- *Prolonged LP and prolonged LP—P37 interval*: This suggests two lesions affecting the peripheral nerve and central conduction. A single lesion of the cauda equina is possible.

Clinical Correlations

Table 16.6 presents the expected SEP findings in various disorders.

Transverse Myelitis

Transverse myelitis produces slowing of SEPs that depends on the site of the lesion. Lesion in the lower cervical or thoracic cord increases central conduction time without having an effect on brain conduction time. With recovery, the SEPs abnormalities are improved, but may not return to normal.

Multiple Sclerosis

SEP is abnormal in most patients with MS, and can be supporting evidence for a silent lesion or confirmatory for a myelopathy. The most common finding in MS is an increase in CCT of the tibial SEP with normal peripheral conduction (LP). This is because the tibial SEP is assessing conduction along the longest myelinated nerve tract of any of the evoked potentials. BCT of median nerve SEP is less commonly increased than tibial nerve SEP CCT. A combined increase in BCT and CCT can be due to tandem lesions, but also can be due to a single lesion in the cervical cord.

Peripheral Neuropathy

Peripheral neuropathy slows peripheral conduction (N9 and LP) with normal BCT and CCT. The N9–P14 interval may be prolonged with lesions affecting the proximal portions of the nerves, such as Guillain–Barré syndrome (GBS). GBS may also occasionally prolong CCT, presumably by affecting the afferent nerve roots of the cauda equina.

Vitamin B_{12} Deficiency

Subacute combined degeneration from vitamin B_{12} deficiency delays or abolishes the cervical and scalp SEPs. With treatment, the abnormalities improve, although not completely to normal. This parallels the clinical course, where there is improvement but also some persistent deficit.

Spinal Cord Injury

Cord transection abolishes potentials above the lesion, but most lesions are incomplete, so the defect in the SEP is variable. Lesions affecting position sense are most likely to alter the SEP. SEP is not perfectly sensitive so many patients may have undetectable scalp potentials despite preservation of some cord function.

Brain Death

Brain death is usually evaluated by EEG or blood flow studies, so the SEP is not typically used as a confirmatory test. In brain death, the scalp potentials are absent, usually with preservation of cervical potentials.

Stroke

SEPs are not commonly used for evaluation of stroke, but if performed, will show attenuation, delay, and often absence of scalp potentials with stimulation of the limbs of the affected side. Lesions of the motor-sensory cortical regions are much more likely to produce abnormal SEPs than lesions elsewhere in the brain. In general, the severity of the stroke deficit correlates with the degree of abnormality of the SEP, but wide variation is common. SEP may be absent with subtle deficit and SEP may be preserved with major deficit.

Part V **Polysomnography**

17 Physiology of Sleep and Sleep Disorders

Polysomnography

Polysomnography (PSG) may be the most underutilized neurophysiologic study. Many patients with sleep disorders are undiagnosed because of lack of suspicion by treating physicians and difficulty getting good-quality sleep studies. Pulmonologists dominate use of the sleep labs in many centers. Neurophysiologists in training should try to gain experience and expertise in PSG so we can play a bigger role in diagnosis and treatment of sleep disorders.

PSG is the recording of EEG and other physiologic measures during sleep. The most common indications for performance of the studies are:

- Excessive daytime sleepiness
- Insomnia
- Limb jerking at night
- Sleep attacks

The most common diagnoses made after sleep-lab testing are:

- Obstructive sleep apnea
- Central sleep apnea
- Narcolepsy
- Periodic limb movements

Sleep stages were discussed in Chapter 4. Some of the information below may be duplicative, but is important to understanding of sleep physiology and pathology.

Physiologic Basis of Sleep

The sleep–wake cycle is controlled by the reticular activating system (RAS). The RAS consists of the brainstem reticular formation, posterior hypothalamus, and basal forebrain. These sites should be conceptualized as a continuous and indistinct structure rather than as separate nuclei.

The exact mechanisms of sleep and wake onset are not known. Activity in the pontine reticular formation, midbrain, and posterior hypothalamus are important for wakefulness. Activity in the medullary reticular formation is important for generation of sleep. Sleep and wake may be integrated in the basal forebrain.

Wakefulness is probably a function of tonic activity in cells that project to the cortex. This activity increases neuronal excitability and may gate reactions to exogenous stimuli. Sleep develops as an active process that is generated in sleep-promoting neurons, such as the serotonergic raphe nuclei. The activation is probably promoted by a reduction in exogenous and endogenous stimuli that indicates a need for sleep. The tonic activating discharge and the response to exogenous stimuli are then suppressed, as are the patterned spontaneous activities normally seen while awake.

Years of sleep deprivation experiments have not explained the need for sleep. One proposed theory promotes the concept of sleep as a time for data management and reorganization. During the waking state, the brain receives a great deal of information on everything from music to tennis to physics. Much of this information is not ordered in a conceptual format, and the brain cannot access the information in a structured way. For example, there is a great difference between owning a tape of a lecture and understanding its content. Some data processing can occur in the waking state, but sleep may be required to organize the day's input, integrate it with existing data, and perhaps discard information that is seldom accessed or is judged by the brain to be useless or uninteresting.

Sleep Stages

Waking State

EEG in the waking state is discussed in Chapter 4 and summarized in Table 17.1. The adult waking EEG consists of predominantly fast frequencies.

TABLE 17.1 Sleep stages

Stage	EEG activity							EMG activity
	Alpha	Beta	Theta	Delta	Vertex	K	Spindle	
Awake	+	+	±	−	−	−	−	+
1A	±	−	+	±	−	−	−	±
1B	±	±	+	−	±	−	−	±
2	−	±	+	+	+	+	+	±
3	−	−	+	+	±	±	±	−
4	−	−	+	+	±	±	±	−
REM	−	+	+	+	−	−	−	−

+ = present; ± = may be present but not prominent; − = not present, or to a minor degree.

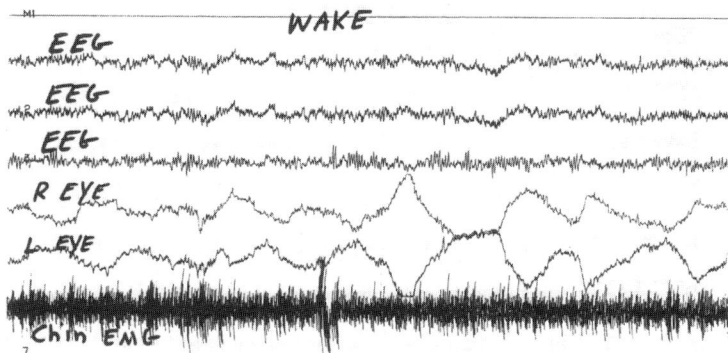

FIGURE 17.1 *Normal waking EEG. Muscle activity and eye movements are evident. There is a paucity of slow activity in the EEG channels.*

When the eyes are closed, a posterior dominant alpha rhythm predominates (Figure 17.1).

Sleep Stage 1
Stage 1 is divided into stage 1A, light drowsiness, and stage 1B, deep drowsiness. Stage 1A shows desynchronization of the background with loss of the posterior alpha rhythm. Theta activity is present but is not prominent (Figure 17.2).

Stage 1B is similar to stage 1A except that slow waves, mainly in the theta range, appear. Vertex waves may be seen during this stage. Positive occipital sharp transients of sleep are seen during this stage.

Sleep Stage 2
Stage 2 is light sleep. For EEG purposes, we do not consider a study to include sleep unless stage 2 is seen. The background consists of a mixture of frequencies. Delta activity is present although not as prominent as in deeper stages of sleep (Figure 17.3). Theta and faster frequencies are superimposed. Differentiation from stage 1B is made by the appearance of sleep spindles. Fusion of sleep spindles with vertex waves results in the K complex. Vertex waves and K complexes are frequent.

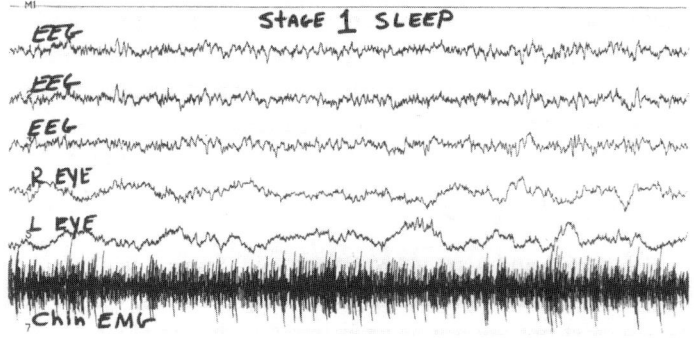

FIGURE 17.2 *Normal stage 1 sleep.*

Physiology of Sleep and Sleep Disorders 235

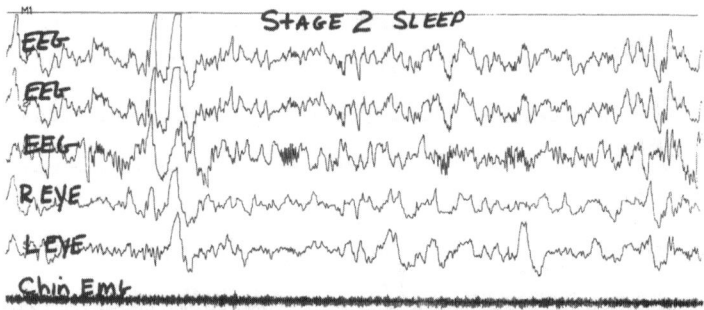

FIGURE 17.3 *Normal stage 2 sleep. Sleep spindles and vertex waves are typical of this stage.*

Stage 3 Sleep
Stages 3 and 4 are slow-wave sleep. Stage 3 is characterized by delta activity with a frontal predominance. Sleep spindles, vertex waves, and K complexes persist but are not as prominent as during stage 2. Mittens are seen during this stage and are composed of a vertex wave fused to the end wave of a spindle. The small spindle wave is the thumb of the mitten and the slow vertex wave is the hand.

Stage 4 Sleep
Stage 4 sleep is characterized by predominance of slow activity in the delta range (Figure 17.4). The delta has a frontal predominance. Although some faster frequencies may be superimposed, sleep spindles and vertex waves are seldom seen and, if present, are poorly formed.

Rapid Eye Movement Sleep
Rapid eye movement (REM) sleep is characterized by the predominance of low-voltage fast activity (Figure 17.5). Eye movement and EMG recordings are especially helpful in differentiating this stage from a normal drowsy pattern. REM sleep usually does not occur within 60 minutes of sleep onset. Sleep-onset REM is seen in patients with narcolepsy, however. Early onset REM may also be seen in some patients with sleep-deprivation, delirium tremens, and brainstem lesions.

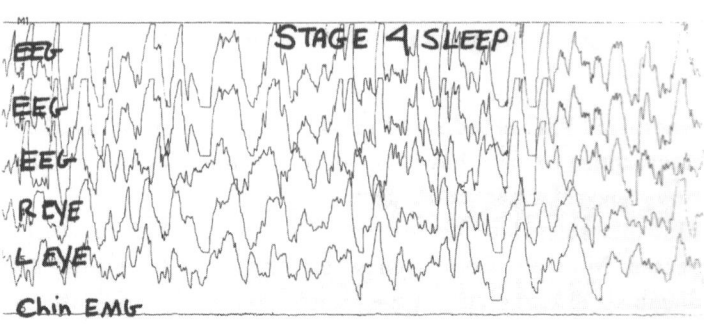

FIGURE 17.4 *Normal stage 4 sleep. Slow activity predominates.*

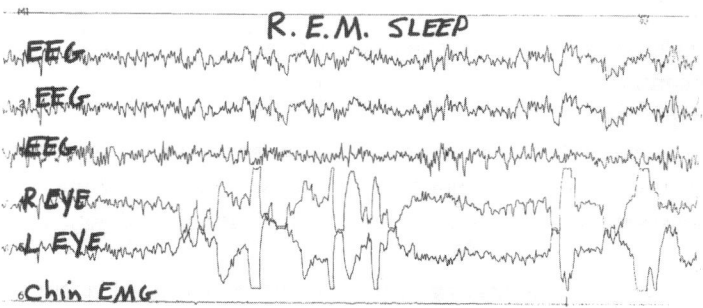

FIGURE 17.5   *Normal REM sleep.*

Indications for Sleep Studies

PSG includes nocturnal sleep testing and multiple sleep latency testing. Ambulatory PSG is also used, though not commonly and will not be discussed further. The *Guidelines* (American Electroencephalographic Society, 1994) make recommendations for clinical indications, but the following are open to personal bias and experience:

- Nocturnal PSG is indicated for patients who have clinical evidence of sleep apnea or who have excessive daytime sleepiness. These conditions suggest a nocturnal sleep disorder.
- A multiple sleep latency test (MSLT) is indicated when narcolepsy is suspected. MSLT testing cannot be used to support the diagnosis of narcolepsy if the clinical history is not consistent. Excessive daytime sleepiness, alone, is not an indication for MSLT.

The *Guidelines* official recommendations list the following indications for sleep monitoring:

- Episodes of sleep at inappropriate times
- Insomnia
- Hypersomnia
- Atypical behavioral events during sleep, for instance somnambulism, seizures, respiratory abnormalities, excessive movements
- Assessment of effectiveness of treatments for sleep disorders

Long-duration EEG monitoring of patients with suspected seizures usually does not require all of the physiologic monitoring commonly performed during PSG studies; the important information is EEG, ECG, and visual monitoring.

18   Sleep Disorders

Basic information is presented for understanding the performance and interpretation of polysomnography (PSG) including nocturnal PSG and multiple sleep latency tests (MSLT). In-depth discussion of methods and interpretation can be found in two excellent texts (Pressman, 2001; Krygen et al., 2000).

The technical recommendations for PSG are taken from the *Guidelines in EEG, Evoked Potentials, and Polysomnography* (American Electroencephalographic Society, 1994). These are referred to as the *Guidelines*.

Nocturnal Polysomnography

Physiologic Measurements

The *Guidelines* recommend that the following physiologic measurements should be made:

- EEG
- Electro-oculogram (EOG)
- Submental EMG
- ECG
- Respiration
- Blood oxygen saturation
- Expired CO_2
- Body and limb movement
- Audiovisual monitoring
- Time

Not all laboratories record all of these parameters, however, the potential for error increases with decreased information. Table 18.1 shows parameters for recording of PSG.

TABLE 18.1 Polysomnography recording parameters

Recording measurement	Electrode montage
EEG	C3-A2
	C4-A1
	O1-A2
	O2-A1
Electro-oculogram	OS-A1
	OD-A2
EMG	Submental leads

Electroencephalography

At least six channels of EEG should be recorded. The following electrode positions should be used, as a minimum:

- Fp1 and Fp2
- C3 and C4
- O1 and O2
- T3 and T4

Electrodes are attached using the same techniques described for routine EEG. Collodion is more dependable for long-term recording than the use of electrode gel alone and is therefore preferable for PSG.

Montages can be determined by the neurophysiologist, but in general should resemble the montages use for routine EEG. Since most PSG equipment is computerized, nowadays, individual channels are recorded and the montage can be selected during review. Versions of the longitudinal bipolar and transverse bipolar montages are used, but only one montage should be used to aid sleep scoring. During interpretation, paper speed is routinely set at 10 mm/sec, although faster speeds are occasionally helpful. With digital recording, this can be changed at the time of interpretation.

Electro-oculogram

Two channels are routinely used for EOG. Electrodes are placed in the following positions:

- 1 cm above and 1 cm lateral to the left eye
- 1 cm below and 1 cm lateral to the right eye
- Left ear or mastoid
- Right ear or mastoid

The two channels are each eye in reference to the ipsilateral ear. Using this montage, eye movements can be clearly differentiated from frontal slow activity. Eye movements will produce potential of opposite polarity in the two eye leads. Frontal slow activity will produce either slow waves of the same polarity or independent slow waves in the two channels.

Submental Electromyography

Submental EMG is recorded using standard cup electrodes placed underneath the chin. The electrodes are connected to a standard EEG amplifier with the low-frequency filter set at 10 Hz and the high-frequency filter set at 70 Hz. Gain is adjusted for each individual but is in the same range as for EEG, about 7 µV/mm.

Submental EMG is reduced in deep stages of sleep and virtually abolished in REM sleep. The absence of EMG activity aids in identification of REM sleep.

Electrocardiogram

ECG is recorded using two electrodes on the chest, usually the rostral sternum and left lateral chest. Self-stick ECG electrodes are satisfactory. The low-frequency filter is set at 5 Hz and the high-frequency filter is set to 70 Hz. Gain is individualized but is usually about 75 µV/mm.

ECG recording serves two purposes. First, heart rate can change with respiratory distress, so that patients with sleep apnea leading to hypoxia and hypercarbia may have initial tachycardia followed by profound bradycardia. In this situation, ECG gives an estimate of the severity of the apnea. The second purpose of ECG recording is to identify cardiac artifact on EEG channels. This is less of a problem for PSG than for EEG performed for epilepsy evaluation. Bipolar montages have less potential for ECG artifact than referential montages.

Respiration

Respiratory monitoring is essential for diagnosis of sleep apnea. Measurements are made of respiratory effort and airflow. Respiratory effort is recorded using thoracic and abdominal strain gauge transducers, intercostal EMG, or thoracic and abdominal impedance. Airflow is usually monitored using thermal sensory near the nares and mouth. DC recordings are preferable, but AC recordings with a long time-constant are acceptable.

Absence of airflow with preserved or enhanced respiratory effort indicates obstructive sleep apnea. Absence of airflow with depression in respiratory effort indicates central sleep apnea. Changes in respiration have to be correlated with blood oxygenation and expired CO_2.

Blood Oxygenation

A pulse oximeter is most commonly used for measurement of oxygen saturation. The output of the oximeter is fed to a recorder by a DC amplifier, since absolute measurements require steady-state DC recordings. The oximeter probe is usually on an earlobe but can be on a finger. Oximeters are fairly accurate, but may give falsely high readings in patients with carbon monoxide in their blood, especially active smokers. Oximeters may give falsely low results in patients with cool extremities or peripheral vascular disease.

Expired CO_2

Expired CO_2 can be measured by placing small sampling tubes below each nostril and near the mouth. The chemical analyzer for CO_2 is rapid and the output can be fed to the recording device. The air at end expiration is largely alveolar, so that determination of CO_2 content is a fairly good indication of gas exchange. Patients with obstructive disorders will have a fall-off in expired CO_2 during the obstruction and may have higher CO_2 content after the obstruction.

Body Movement

Surface EMG electrodes are placed over the tibialis anterior on one side for recording EMG. This can reveal the presence of myoclonus and may aid in the diagnosis of restless legs syndrome. Alternatively, accelerometers can be used. These are small devices that produce a signal with a small amount of movement. However, because submental EMG is already recorded, another EMG channel is much more convenient to use.

Audiovisual Monitoring

Closed-circuit television (CCT) and microphone recording are used for monitoring of behavior and movements. If the recording is not digital, split-screen recording of the EEG and video signal is desirable; however, on digital systems, the streaming media is cued to the EEG and physiologic display.

The camera is placed so that it can easily record the patient in the bed. Modern cameras are able to record in the very low-light conditions that are conducive to sleep. Additional light may be provided from an infrared source. These light sources activate detectors in the camera but will not wake the patient. Audio monitoring by small microphones can detect vocalizations and be another indicator of respiratory effort. The most important aspects of audiovisual monitoring are muscle twitches, signs of arousal, axial and limb movements, respiratory effort, and seizures.

Time

Time is measured by a real-time clock in the computer, and is expressed as real and elapsed time. Time of onset and cessation of recording should be noted on the record. Paper records need to have accurate indicators of time, to associate with the audiovisual monitoring. Unfortunately, comparing paper recordings and videotape monitoring is much more difficult than simultaneous viewing on digital systems, and there is greater potential for error in timing with the paper records.

Recording Protocol for a Standard Nocturnal Study

The following guidelines summarize the recommendations for a standard nocturnal PSG recording:

- Make the room comfortable and quiet. Place the recording equipment in a separate room.
- Begin the study as close to normal sleep time as possible.
- Minimize interruptions. Extra electrodes and sensors facilitate maintaining adequate recording if the patient dislodges primary electrodes and sensors.
- Duration of recording should ideally be 8 hours, with 6.5 hours as a minimum.

Interpretation

Grading of nocturnal PSG is typically done in 20- to 40-second epochs. An epoch is classified according to the predominant pattern during the epoch. For example, an epoch characterized mainly by a desynchronized background may be classified as stage 1 sleep even though there is some occasional posterior waking alpha activity.

Sleep onset is defined as either the first of three contiguous epochs of stage 1 sleep or the first of any stage 2, 3, or 4 sleep. Three consecutive epochs of state 2, 3, or 4 sleep are not required. With simultaneous monitoring of many physiologic variables, the amount of generated data can be overwhelming. The *Guidelines* recommends the following sleep measurements.

- Total time in bed
- Duration of interspersed wakefulness
- Total sleep time
- Sleep latency
- REM latency
- Number of awakenings
- Time in each sleep stage (actual and percentage)

Sleep efficiency is the percentage of total time in bed spent asleep. In addition, graphs of sleep stage progression are drawn. These are usually generated automatically by the acquisition software.

Figure 18.1 shows a histogram of sleep stages in a normal patient. The following measurements are made:

- Respiratory rate in the wake and sleep stage
- Presence of snoring
- Presence of paradoxical respiratory patterns
- Number and type of apneic episodes
- Frequency and degree of oxygen desaturation

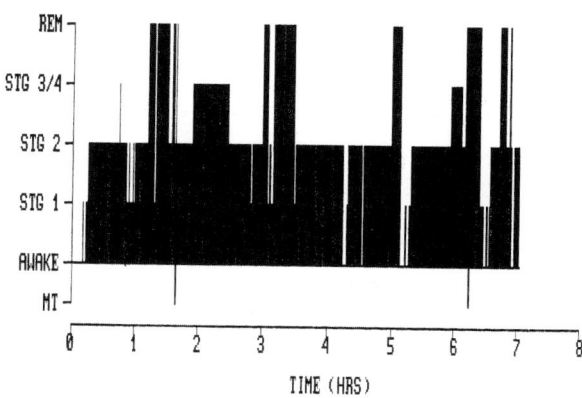

FIGURE 18.1 *Histogram of sleep stages showing periodic awakenings interspersed with the sleep stage progressions.*

ECG data are analyzed for the following:

- Mean and range of heart rate during wake and sleep states
- Arrhythmias, if present
- Cardiac response to respiratory changes, for instance apnea

EMG data are analyzed for myoclonus and differentiation is made between myoclonus associated with arousal, myoclonus associated with epileptiform activity on EEG, and myoclonus not associated with other physiologic changes.

The data obtained from PSG is voluminous. Some of the important sleep parametric data obtained includes: total sleep time, sleep efficiency, wake after sleep onset, wake after sleep offset, sleep latency, REM latency, and percentages of sleep in each sleep stage. While there are other data, these are the most important.

Multiple Sleep Latency Test

The multiple sleep latency test (MSLT) is performed in the daytime and can be performed in standard EEG labs without all of the monitoring equipment required for nocturnal PSG. Patient preparation is minimal, although it is important to ensure that no sedatives are taken within 1 week of the test, since the results will be influenced by sedative drug effect or sedative withdrawal. The *Guidelines* recommends performing the MSLT during the day following a nocturnal sleep study, so that the quality of sleep is known.

Methods

Conventional EEG electrodes are used, and a waking recording is made. The patient is then asked to go to sleep. The technician marks the time on the record.

The MSLT is performed as follows:

1. The patient has a normal night's sleep prior to the recording. Most neurophysiologists believe that it is important to have the patient under PSG study to evaluate the quality of the previous night's sleep. This helps to determine whether a positive MSLT might be caused by a disorder of nocturnal sleep, rather than being primary. A PSG recording is probably not necessary in all patients.
2. Electrodes are placed according to the 10–20 Electrode Placement System. The entire array is probably not necessary; however, it is easily placed in most EEG laboratories. If a limited array is placed, central and occipital leads are essential for identification of central vertex activity and the posterior dominant rhythm. In addition to EEG leads, electrodes should be placed for monitoring the following physiologic parameters:
   a. EOG
   b. Submental EMG
   c. ECG

3. At least four naps are begun at scheduled intervals. The technician lowers the lights and asks the patient to go to sleep. Approximately 15 minutes are recorded before the first nap. After the "goodnight" command, recordings are made until the following criteria are met:
   a. 20 minutes without sleep
   b. 15 minutes of continuous sleep
   c. 20 minutes of interrupted sleep, even if less than 15 minutes of sleep occurred

Interpretation

The following physiologic measurements are made during the MSLT:

- Latency from "goodnight" to sleep onset
- Latency from sleep onset to REM sleep

Forty-second epochs are scored according to the predominant background. If an epoch is mainly desynchronized with slow roving eye movements but contains a small amount of posterior dominant alpha, the epoch I still scored as stage 1. Sleep onset is identified as the first of three consecutive stage 1 epochs or any epoch of stage 2, 3, 4, or REM sleep.

Mean sleep latency is the average of the sleep latencies determined for each nap. Some neurophysiologists do not consider a test to be interpretable unless the patient falls asleep with two or three of the naps.

The most characteristic abnormality found with sleep latency tests is short sleep latency. A mean sleep latency of less than 5 minutes is virtually diagnostic of hypersomnolence. Latency of 10 minutes or more is normal. A mean sleep latency between 5 and 10 minutes is borderline. The report may indicate that consistent mean sleep latency of less than 10 minutes is suggestive of a sleep disorder but is not diagnostic.

Mean sleep latency is altered by the following conditions:

- Sleep deprivation
- Certain medications, especially sedatives, antihistamines, and stimulants
- Withdrawal of some medications
- Age

Withdrawal of medications such as benzodiazepines and barbiturates can have sustained effects, therefore, the medications should be stopped at least 2 weeks before the study.

Many patients with excessive daytime sleepiness will have shorter sleep latencies. Patients with narcolepsy will often have sleep-onset REM periods. The interpreter should be sure that the patient is not sleep-deprived before making the conclusion of sleep-onset or short-latency REM periods.

TABLE 18.2　Sleep disorders

Disorder	Findings
Narcolepsy	REM-onset sleep
Obstructive sleep apnea	Respiratory effort without air movement Eventual arousal
Central sleep apnea	Loss of respiratory effort during drowsiness or deep sleep Eventual arousal
Drug-related insomnia	Fragmented REM-sleep periods

REM = rapid eye movement.

Disorders

Sleep disorders can be classified according to the following categories:

- Hypersomnias
- Insomnias
- Disorders of the sleep–wake cycle
- Arousal and paroxysmal disorders in sleep
- Excessive daytime sleepiness

Some neurophysiologists include seizures evoked by sleep as a sleep disorder. This is really an underlying seizure disorder with sleep as an activating method. The neurophysiologic findings in specific sleep disorders are summarized in Table 18.2.

Narcolepsy

Narcolepsy is characterized by daytime sleep attacks. There are two basic types of narcolepsy:

- Non-REM, or isolated, narcolepsy
- REM, or compound, narcolepsy

Non-REM narcolepsy has non-REM sleep during attacks. Night sleep is normal. REM-narcolepsy has REM sleep during sleep attacks. At night, there is sleep fragmentation and shortened REM latency.

The MSLT may detect short REM latency or sleep-onset REM. The MSLT tests naps but may catch a narcoleptic sleep attack.

Sleep Apnea

Sleep apnea is probably the most common clinical reason for ordering PSG. There are three basic types of sleep apnea:

- Obstructive sleep apnea (OSA)
- Central or nonobstructive sleep apnea
- Mixed sleep apnea

All types are characterized by loss of airflow for 10 seconds or more. Patients with OSA continue to have respiratory effort that gradually increases because the movements are ineffectual. Eventually, partial arousal results in opening of upper airway passages and restoration of ventilation. Patients with central sleep apnea lose air movement because of loss of respiratory drive. With subsequent hypoxia and hypercarbia, there is partial arousal and restoration of normal ventilation.

Appendix: Abbreviations and Units of Measurement

ACh	acetylcholine
AChE	acetylcholinesterase
AChR	acetylcholine receptor
ADC	analog-to-digital converter
AIDP	acute inflammatory demyelinating polyradiculoneuropathy
ALS	amyotrophic lateral sclerosis
APB	abductor pollicis brevis
ATP	adenosine triphosphate
BAEP	brainstem auditory-evoked potential
C	symbol for capacitor
Ca^{2+}	chemical abbreviation for calcium
CD	compact disk
CIDP	chronic inflammatory demyelinating polyneuropathy
Cl^-	chemical abbreviation for chloride
CMAP	compound motor action potential
CMT	Charcot–Marie–Tooth (disease)
CNS	central nervous system
CRT	cathode-ray tube
CTS	carpal tunnel syndrome
DL	distal latency
DTR	deep tendon reflexes
ECG	electrocardiography (-gram)
EDC	extensor digitorum communis
EEG	electroencephalography (-gram)
EMF	electromotive force
EMG	electromyography (-gram); usually implies the needle study, although surface and single-fiber studies are also included in this term
EP	evoked potential
EPSP	excitatory postsynaptic potential
ER	emergency room
FIRDA	frontal intermittent rhythmic delta activity
G	symbol for conductance
HFF	high-frequency filter
HMSN	hereditary motor sensory neuropathy
ICU	intensive care unit
IPSP	inhibitory postsynaptic potential

K+	chemical abbreviation for potassium
LB	longitudinal bipolar (montage)
LEMS	Lambert–Eaton myasthenic syndrome
LFF	low-frequency filter
MEP	motor-evoked potential
MEPP	miniature end-plate potential
MG	myasthenia gravis
MMN	multifocal motor neuropathy
MUP	motor unit potential
MS	multiple sclerosis
Na+	chemical abbreviation for sodium
NCS	nerve conduction study (includes assessment of nerve conduction velocity, amplitude, and waveform)
NCV	nerve conduction velocity
NMJ	neuromuscular junction
OIRDA	occipital intermittent rhythmic delta activity
OR	operative room
PDR	posterior dominant rhythm
PDS	paroxysmal depolarization shift
PIRDA	posterior intermittent rhythmic delta activity
PLED	periodic lateralized epileptiform discharge
PNS	peripheral nervous system
POSTs	positive occipital sharp transients
Ref	reference (electrode or montage)
REM	rapid eye movement (sleep)
SEP	somatosensory-evoked potential
SNAP	sensory nerve action potential
SPL	sound pressure level
SSPE	subacute sclerosing panencephalitis
TA	tibialis anterior
TB	transverse bipolar (montage)
TC	time constant
TCD	transcranial doppler
TM	transverse myelitis
VEP	visual-evoked potential

Units of Measurement

cm	centimeter
dB	decibel
Hz	hertz
kHz	kilohertz
kohm	kilo-ohm
m	meter
µV	microvolt
µs	microsecond
mm	millimeter
ms	millisecond
mV	millivolt
sec	second
V	volt

Glossary

Action potential
Regenerative membrane potential produced by activation of voltage-dependent sodium channels.

Active filter
Filter that uses energy to accomplish its task, as opposed to a passive filter.

After-hyperpolarization
Increased membrane potential that follows an action potential, associated with a refractory period.

Aliasing
Alteration in waveform due to digital sampling at too slow a rate. Some frequencies may be mistaken for slower frequencies.

Alpha
EEG activity in the 8- to 12-Hz range.

Alpha coma
Coma due to severe brain damage, where there is diffuse invariant alpha activity. In general, this pattern signifies a poor prognosis.

Ampere
A unit of current. One coulomb of charge flowing past a point in a conductor each second is a current of one ampere, or 1 amp (from André Ampère, a nineteenth-century French physicist).

Amplifier
Device that increases the size of a signal.

Analog data
Data represented as a continuous variable quantity that is independent of the scale used to measure the value. Compare with *digital data*.

Analog-to-digital converter
Device that converts analog data to digital data using a defined sampling rate and voltage resolution.

Anion
Negatively charged ion.

Anode
Positive terminal of a power supply.

Antidromic
Stimulation of a nerve so that the action potential is conducted in the opposite direction to normal nerve impulse flow. Opposite of *orthodrodromic*.

Axonal neuropathy
Damage to peripheral nerve where the predominant pathology is to the axon, producing denervation on EMG and reduced compound motor and sensory nerve action potential amplitudes in nerve conduction studies. Little or no slowing of the nerve conduction velocities.

Axonotmesis
In nerve trauma, breakage of the axons with the connective tissue sheath remaining intact.

Beta
EEG activity above 12 Hz.

Blink reflex
Nerve conduction study procedure that is a brainstem reflex used to assess facial and trigeminal nerve function and brainstem reflex pathways.

Breach rhythm
EEG pattern due to a skull defect, where there is a localized prominence of beta activity.

BSAPPs
Brief small-amplitude polyphasic potentials, also called *myopathic motor unit potentials*.

Capacitor
Device that stores energy by separation of charge.

Cathode
Negative terminal of a power supply.

Cation
Positively charged ion.

Complex repetitive discharge
Repetitive discharge of several muscle fibers, considered abnormal but not a specific finding.

Conduction block
Segmental slowing of block of nerve conduction.

Coulomb
A unit of measurement of charge, equivalent to the charge carried by 6.24×10^{18} electrons (from Charles de Coulomb, an eighteenth-century French physicist).

Current
Flow of charge. Current is described as the flow of positive charge, which is opposite to the direction of flow of electrons.

Decibel (dB)
A unit of measure of electrical signal intensity or sound intensity. For electrical signals, the numbers of decibels difference between two signals is equal to 20 times the log of the ratio of the amplitudes of the signals. For example, a thousand-fold difference in signal intensity is a 60-dB difference. To make things confusing, for sound, a decibel is 10 times the log ratio of the amplitude of the signals.

Delta
EEG activity in the frequency of less than 4 Hz.

Demyelinating neuropathy
Damage to the myelin sheath that affects conduction of the axons; the axons may fail to conduct at all. Seen in nerve conduction studies as slowing of the nerve conduction velocities with little or no denervation seen on EMG.

Differential amplifier
Amplifier whose output is the amplified difference between the two inputs.

Digital data
Data represented as a discrete measurement. The measurement depends on the scale used to make the measurement.

Diode
Semiconductor device that allows for current flow in only one direction.

Dwell time
Sampling time of an analog-to-digital converter. Archaic term.

Early recruitment
Abnormal EMG pattern where voluntary contraction evokes activation of motor unit potentials at a lower level of effort than normal, due to ineffective contraction of myopathic muscle fibers.

Electromotive force
Force that controls the flow of charged particles, especially electrons.

Electron
Negatively charged atomic particle.

Electrotonic conduction
Passive conduction of a potential difference down a membrane.

Equilibrium potential
Membrane potential for a specific ion that exactly counteracts the chemical gradient of the ion to move.

Excitatory postsynaptic potential
Postsynaptic potential produced by release of an excitatory transmitter.

F wave
Response recorded from motor nerves in response to antidromic stimulation of the same motor nerve. The response is reflected off the motor neuron in the spinal cord.

Fibrillations
Spontaneous discharge of muscle fibers, in acute denervation and in myopathies.

Filter
Circuit device that alters the frequency composition of the signal.

Generator potential
Potential generated by membrane activation.

H reflex
Electrophysiologic equivalent of the tendon reflex, where stimulation of the afferent fibers of a motor nerve results in activation of the efferent motor axons.

Inductor
Device that stores energy in the form of a magnetic field.

Inhibitory postsynaptic potential
Postsynaptic potential produced by release of an inhibitory transmitter.

Leak current
Current that flows from a circuit to ground. This is usually a small amount of current but can be potentially dangerous.

M response
The response recorded from a muscle when a motor nerve is stimulated. Equivalent to the compound motor action potential.

Montage
Connection of electrode channels in EEG.

Mu rhythm
Central 10- to 12-Hz rhythm present in the waking state that is abolished by moving the contralateral hand. Normal pattern.

Myopathic motor unit potential
Potential that signifies myopathy—usually short-duration polyphasic potential.

Myopathy
Disorder of muscle.

Myotonia
Repetitive muscle fiber activation potentials seen on EMG in myotonic dystrophy, myotonia congenita, paramyotonia congenita, and hyperkalemic periodic paralysis.

Neurapraxia
In nerve trauma, disruption of nerve fiber function without transection of the axons.

Neuronopathy
Damage to the neuronal cell body, such as with amyotrophic lateral sclerosis.

Neuropathic motor unit potential
High-amplitude, long-duration polyphasic motor unit potential. Abnormal EMG pattern.

Neuropathy
Disorder of peripheral nerve.

Neurotmesis
In nerve trauma, disruption of nerve fiber function without transection of the axons.

Neutron
Neutral atomic particle.

Ohm
The unit of resistance. A 1-ohm resistance allows 1 amp of current to flow from a 1-V battery (from Georg Ohm, a nineteenth-century German physicist).

Orthodromic
Stimulation of a nerve so that conduction of the action potential is in the same direction as normal nerve impulses. Opposite of *antidromic*.

Passive filter
Filter that does not use exogenous energy.

Photon
Packet of light.

Polymorphic delta activity
Slow activity with an irregular configuration.

Positive sharp waves
Spontaneous discharge of muscle fibers, seen in active denervation and in myopathies.

Posterior slow waves of youth
Slow waves superimposed on the waking EEG background. Normal pattern.

Potential difference
The amount of energy required to separate a given quantity of charge. This is a measure of the ability to do work, measured in volts.

Power
The rate at which energy can be transferred from a power supply to electrons. Also, the square of the amplitude of a specific frequency found after frequency analysis of a waveform.

Proton
Positive atomic particle.

Pseudomyotonic discharge
Archaic term for *complex repetitive discharge*.

Reduced recruitment
Abnormal EMG pattern with voluntary contraction, where there are reduced numbers of available motor units.

Refractory period
Period after an action potential when the membrane cannot be excited to produce an action potential.

Resistor
Circuit device that dissipates the energy associated with moving electrons as heat.

Resistor–capacitor (RC) circuit
Simplest filter, though not practical.

Sampling interval
Interval between samples measured by an analog-to-digital converter.

Sampling rate
Rate of sampling of an analog-to-digital converter. Sampling rate is the inverse of *sampling interval*.

Secondary demyelination
Damage to the myelin sheath due not to direct myelin damage but to primary damage to the neuron or axon. Seen in nerve conduction studies as mild slowing of conduction velocity when the predominant findings are of axonal damage.

Semiconductor
Material that conducts better than a nonconductor but less well than a conductor.

Theta
EEG activity in the 4- to 7-Hz frequency band.

Time constant
Time that describes the charging and discharging of the capacitor in an RC circuit.

Time-dependent channel
Ionic channel in biological membranes that closes after a specified time.

Transistor
Semiconductor device integral to amplifiers. From the terms transfer and resistor.

Valence
Combining power of an atom—usually related to the electron configuration of the outer orbital.

Volt
A measure of electrical potential difference between two points. This is the driving force for movement of electrons. One volt can drive 1 coulomb of charge per second through a resistance of 1 ohm (from Count Alessandro Volta, an eighteenth–century Italian physicist).

Voltage-gated channel
Ionic channel that is activated by the membrane potential crossing a specified threshold.

Voltage resolution
A measure of the ability of an analog-to-digital converter to discern voltage differences. Determined by the number of bits of the converter.

References

American Encephalographic Society. Guidelines in EEG, evoked potentials, and polysomnography. J Clin Neurophysiol 1994; 11:2–127

Aminoff MJ; Electrodiagnosis in Clinical Neurology (4th ed). Churchill Livingstone, 1999. *Fine comprehensive text. Has good detail on studies and clinical correlations.*

Chiappa KH. Evoked Potentials in Clinical Medicine. Lippincott, 1997. *Excellent text. Especially strong on clinical correlations.*

Fisch BJ; Fisch & Spehlmann's EEG Primer. Elsevier, 2002. *Good basic EEG text. Extensively updated from the earlier editions.*

Hughes JR; EEG in Clinical Practice. Butterworth-Heinemann, 1994. *Terrific book which is a good read for the student of EEG. Hopefully will be updated, but still useful as is.*

Kimura J; Electrodiagnosis in Diseases of Nerve and Muscle. Oxford, 2001. *Wonderful comprehensive EMG text. Especially helpful for the experienced EMGer.*

Krygen MH, Roth T, Dement WC; Principles and Practice of Sleep Medicine. Saunders, 2000. *Concise and useful sleep text, helpful for reading cover to cover as well as a reference.*

Misulis KE, Fakhoury T; Spehlmann's Evoked Potential Primer. Butterworth-Heinemann, 2001. *Update of a good basic EP text.*

Misulis KE, Fenichel GM; Genetic forms of myasthenia gravis. Pediatr Neurol 1989; 5:205

Osselton JW et al.; Clinical Neurophysiology: Electromyography, Nerve Conduction and Evoked Potentials. Butterworth-Heinemann, 1995. *Good text with good methodological detail. Helpful for those new to EMG as well as experienced practitioners.*

Perotto A; Anatomical Guide for the Electromyographer: Limbs and Trunk (3rd ed). Thomas, 1994 *Currently out of print, but an excellent reference. Might be found used on the Internet.*

Pressman M; Primer of Polysomnogram Interpretation (2nd ed). Butterworth-Heinemann, 2001. *Great basic text with fine clinical correlations to findings. Easy to read.*

Preston DC, Shapiro BE; *Electromyography and Neuromuscular Disorders: Clinical–Neurophysiologic Correlations.* Butterworth-Heinemann, 1998. *Terrific text with lots of explanation and discussion with clinical correlations. Graphics and tables are the best of these texts.*

Report of the Medical Consultants on the Diagnosis of Death to the President's Commission for the Study of Ethical Problems in Medicine and Biomedical and Behavioral Research. Guidelines for the determination of death. JAMA 1981 (Nov. 13); 246(19): 2184–6

Russell GB; Primer of Intraoperative Evoked Potential Monitoring. Butterworth-Heinemann 1995. *Good solid text, more information on clinical correlations.*

Task Force for the Determination of Brain Death in Children. Guidelines for the determination of brain death in children. Arch Neurol 1987; 44:587

Index

Abductor pollicis brevis (APB), 173
Absence seizures
   EEG activation, 59, 99
   spikes, 88, 99
Acetylcholine (NMJ), 10, 130, 133, 151
   autoantibodies, 187
   breakdown, 190–191
   miniature end-plate potentials (MEPPs), 152
Acetylcholinesterase deficiency, 190
Acoustic neuroma, BAEPs, 218
Action potentials, 7–8
   afferent, 131
   definition, 7
   efferent, 41, 130
   fibrillation potentials, 132
   F-waves, 141
   ionic mechanisms, 7
   motoneurons, 130
   phases, 7
      after-hyperpolarization, 8
      generator potential, 7
      peak potential, 8
      regenerative depolarization, 7
      repolarization, 8
   propagation, 8–9
      demyelinated axons, 133
      myelinated vs. unmyelinated axons, 8
   refractory periods, 8
Activation methods (EEG), 58–59, 81–84
Active filters, 33
Active sleep, neonatal EEG, 109, 111
Acute inflammatory demyelinating polyradiculoneuropathy (AIDP). *See* Guillain–Barré syndrome
Adrenoleukodystrophy, SEPs, 228
Afferent nerve volley, SEPs, 221

After-hyperpolarization, 8
Age effects
   alpha rhythm, 66
   sensory conduction studies, 140–141
   wicket spikes, 73
Alcohol, REM-onset sleep, 68
Alpha rhythm, 43–44. *See also* Posterior dominant rhythm (PDR)
   abnormal, 43
      dementia, 65
      encephalopathy, 65
      neonatal, 114
      slow (pathological), 65, 85, 90–91
      spikes, 89
   age effects, 66
   asymmetry, 44, 65
   children, 44, 69
   frequency range, 43, 65
   location, 44
   maturation, 69
   slow variant, 71, 73–74
   suppression, 44, 66, 76–77
Ambulatory EEG monitoring, 122–123
American Electroencephalographic Society *Guidelines*
   auditory-evoked potentials, 212–213
   montage, 54–55
   neonatal EEG, 107
   neonatal EP, 199
   polysomnography, 237, 239, 243, 244
Ampere (amp), 17
Amplifiers, 36–37
   calibration, 56–57
   common mode rejection, 37, 59
   differential, 36–37
Amyotrophic lateral sclerosis (ALS), 131, 168
   NCS/EMG, 164
   SEPs, 228

Analog filters, 47
Analog-to-digital conversion (ADC), 34, 123
Anesthesia, alpha rhythm, 44
Anoxic encephalopathy, EEG, 88, 103
Arbovirus encephalitis, EEG spikes, 88
Arc distortion, 48
Arsenic, 22
Artifacts
   electroencephalography (EEG), 60, 75–81, 84, 120
   evoked potentials (EPs), 197–198
   nerve conduction studies (NCS), 136
Asymmetrical EEG
   abnormal patterns, 113
   alpha rhythm, 44, 65
   beta rhythm, 44
Atomic structure, 15, 16
Audiometry, 218, 219–220
Audiovisual monitoring, nocturnal polysomnography, 242
Auditory-evoked potentials (brainstem). *See* Brainstem auditory-evoked potentials (BAEPs)
Autoimmune disease
   demyelinating, 132–133
   myasthenic, 159, 164, 187–188, 190–191
Autonomic neuropathy, 170
Averaging, 34
   evoked potentials (EPs), 197
   sensory conduction studies, 139
Axonal degeneration. *See* Neuropathies

Bandwidth, definition, 31
Basal forebrain, sleep role, 234
Becker's muscular dystrophy, 181–182
Behavioral disturbances, 97
Bell's palsy, 174
Benign epileptiform transients of sleep (BETS), 73, 89
Benign focal epilepsies of childhood, 88, 96–97
Beta rhythm, 44
   hyperthyroidism, 44
   sedative effects, 44, 65
   waking state, 65
BETS, 73, 89
Biological calibration (biocal), 57
Biopsy, muscle, 184
Bipolar montage, 53, 54, 55
Blink reflex, 129, 142–144
   components, 142
   electrode placement, 143
   interpretation, 143, 144
   parameters, 143
Blood oxygenation, nocturnal polysomnography, 241
Body temperature
   brain death and, 117, 118
   nerve conduction and, 140–141
Botulism, 159, 189–190
Brachial plexitis, 177
Brachial plexus lesions
   NCS/EMG, 162
   plexopathy, 177
   SEPs, 228
Brain conduction time (BCT), 224
Brain death
   children, 119, 120
   clinical examination, 117
   criteria, 118
   EEG, 117–120
      isoelectric, 113
      montage, 119
      technical standards, 120
   evoked potentials
      BAEPs, 219
      SEPs, 228, 230
   guidelines for determining, 117–119
   President's Commission, 117
   Task Force for Brain Death in Children, 119
   transcranial doppler (TCD), 118
Brain mapping, 124–125
   dementia, 124, 125
   epileptiform activity, 46, 124, 125
Brain SEPs, 221
Brainstem
   BAEPs, 217, 218
   SEPs, 228
   sleep role, 234
   stroke, 228
   tumor, 218
Brainstem auditory-evoked potentials (BAEPs), 201, 211–220
   abnormal, 216
   analysis time, 198
   audiometry, 216, 219–220
   clinical correlations, 195, 218–219
   clinical indications, 195
   data analysis, 215–216
      latency, 215
   earphones, 211
   interpretation, 213–216
   methods, 211–213
   neural generators, 196
   normal, 215

pediatric, 216–218
recording, 213
  parameters, 214
stimulus, 211–213
  clicks/tones/white noise, 212
  intensity, 212–213
  parameters, 214
  rate, 212
waveforms
  identification, 213–215
  latency changes, 215–216
  waves I–IV, 213, 214
Breach rhythm, 44
Brief small-amplitude polyphasic potentials (BSAPPs), 133
Burst-suppression pattern, 103, 104
  interpretation, 98
  neonatal EEG, 113

Calcium
  neurodegeneration, 132
  neurotransmitter release, 9
Calibration, EEG, 56–57
Capacitance, 19
  additive, 31
  electrode, 35
  mathematical derivation, 20
  measurement, 19
  stray, 78
Capacitative current, 19
Capacitors, 19–20
  circuit properties, 30–31
  inductors vs., 21
  parallel/series, 30, 31
  resistor–capacitor circuits, 32–33
Carbon dioxide expiration, nocturnal polysomnography, 241–242
Carnitine palmityl transferase (CPT) deficiency, 184–185
Carpal tunnel syndrome (CTS), 153–154, 173–175
  clinical evaluation, 162, 173–175
  common findings, 164
Cathode-ray tubes, 47
Central sleep apnea, 246–247
Cerebral cortex, 41–42
Cervical radiculopathy
  NCS/EMG, 162
  SEPs, 228
Charcot–Marie–Tooth disease
  NCS/EMG, 170, 172
  SEP, 228
Checkerboard patterns, 202

Children
  brain death determination, 119
  EEG, 69–71
    alpha rhythm, 44, 69
    benign focal epilepsies, 88, 96–97
    beta rhythm, 44
    brain death studies, 119, 120
    delta rhythm, 45, 93
    hyperventilation, 84
    maturation of posterior rhythm, 69
    neonates, 107–116
    posterior slow waves of youth, 69
    sensitivity, 58
    sleeping, 69–71
    waking, 69
  evoked potentials
    BAEPs, 216–218
    maturation, 199
    neonates, 199, 218, 219
Chronic inflammatory demyelinating polyneuropathy (CIDP), 133, 171
Circuit elements, 18–25
Circuit laws, 25–28
Circuit loop, 27
Circuit properties, 28–31
Circuit theory, 18
  capacitors, 20
  diodes, 23
  inductors, 22
  passive vs. active circuits, 24
  resistors, 19
  semiconductors, 23
  transistors, 25
Clicks, auditory-evoked potentials, 212
Clostridium botulinum, 189
CMAP. *See* Compound motor action potential (CMAP)
Colloidion fixation of electrodes, 50, 51
Coma
  alpha rhythm, 44
  BAEPs, 219
Common-mode rejection, 37, 59
Common-mode rejection ratio (CMRR), 37, 59
Complex partial seizures, EEG spikes, 88, 96
Complex repetitive discharge (CRD), 149, 150
Compound motor action potential (CMAP), 132, 136, 221
  ALS findings, 168
  H-reflex, 142

Compound motor action potential
  (CMAP)—*continued*
  increased distal latency, 137
  measures, 137
  repetitive stimulation of NMJ, 152,
    188, 189
  waveform abnormalities, 138
Conceptual age, 109
Conductance, definition, 5
Conduction, nerves. *See* Nerve
  conduction
Conduction block, 133, 138, 167
Conductive hearing loss, 219
Conductors, 15, 16
  atomic structure, 16
  magnetic fields, 21
  semiconductor manufacture, 22
Constant-current stimulators, 136
Cortical blindness, 209
Cortical efferents, 41
Cortical potentials, 41–42
Coulomb, 17
Cranial nerves, blink reflex, 142–144
Creutzfeldt–Jakob disease (CJD)
  EEG, 104–105
    spikes, 88
  SEPs, 228
Critical illness polyneuropathy, 162
Current, 16, 17–18
  AC to DC conversion, 24
  capacitative, 19
  impedance, 25
  mathematical derivation, 18
  measurement, 17
  resistance, 18–19
  transistor amplification, 25
Cushing's syndrome (steroid myopathy),
  184

Damping, 48, 57
Dejerine–Sottas disease, 170, 172
Delta brushes, 111
Delta rhythm, 45
  FIRDA, 45, 85, 93–94
  neonates, 111
  PIRDA, 45, 93–94
  polymorphic delta activity (PDA),
    92–93, 94
  sleep, 45, 92
  slowing, 92, 105
  waking state, 92
Dementia
  brain mapping, 124, 125
  slow alpha rhythm, 65

slow PDR, 91
Demyelinating disorders, 132–133
  NCS/EMG, 162, 171
Dendritic arborization, 11, 41
Depolarization
  electronic decay, 6
  field potentials, 12–14
  generator potentials, 6
  motoneurons, 130
  regenerative, 7
  transmitter release, 9
Depth electrodes, 51
Dermatomyositis, 183
Diabetes mellitus
  neuropathy (NCS/EMG), 167,
    170–171, 179
  SEP, 228
Differential amplifiers, 36–37
Digital EEG
  analog vs., 47–49
  display montage, 48
  face sheets, 56
  gain, 48
  mechanical distortion, 48–49
  quantitative EEG, 123–125
    brain mapping, 124–125
    power spectral analysis, 124
    spike detection, 124
  storage, 48, 63
Digital filtering, 34
Digital filters, 33, 47
Digital signal analysis, 34
Diodes, 22–24
  forward/reverse-biased, 23, 24
  rectifiers, 24
Dipoles, 42
Disk electrodes, gel, 35
Dispersion, 133
Distortion, 48–49
Doping, 21, 22
Driving response, photic
  stimulation, 82
Drowsiness (sleep stages 1A/B), 66, 67,
  68, 235
  rhythmic mid-temporal theta of
    drowsiness, 71, 74–75
Drug-induced REM-onset
  sleep, 68
Drug intoxication vs. brain
  death, 117
Duchenne muscular dystrophy,
  181–182
Dysarthria, 162
Dysphagia, 162

Ear, audiometry, 219–220
Earphones, auditory-evoked potentials, 211
Electrical gradients, 4
Electric fields, 17
Electrocardiography (ECG)
   EEG artifact, 80, 120, 241
   neonatal EEG and, 107
   nocturnal polysomnography, 241, 244
Electrode caps, 60
Electrode–gel interface, 79
Electrodes
   active, 35
   capacitance, 35
   colloidion fixation, 50, 51
   disk electrodes, 35
   ECG, 241
   EEG, 50–53
      depth, 51
      ear, 76, 80, 240
      error, 60
      head measurement, 53
      impedance, 59
      inpatient monitoring, 121–122
      leads, 76, 78–79
      magnetic fields, 78
      movement artifact, 78–79
      name assignment, 52, 53
      neonatal EEG, 107
      positioning, 52–53, 60, 76, 240
      removal, 51
      scalp recording, 11–12, 41
      sphenoidal, 51
      subdural strip, 51
      surface, 50–51
   electrode–amplifier interface, 36
   EMG, 144
      insertion, 145
      needle, 129, 145, 146
      positioning, 145, 146
   EP
      BAEPs, 213
      positioning, 204–205, 213, 222, 225
      SEPs, 222, 225
      VEPs, 204–205
   gel fixation, 35, 50
      interference effects, 79
      pulse artifacts, 80–81
   near-field, 12
   needle, 35, 51, 144
   nerve conduction studies (NCS), 135, 138
      blink reflex studies, 142, 143
      F-wave studies, 141
      positioning, 135, 138–139, 143
      surface, 135
   patient–equipment interface, 35–36
   reference, 13, 35, 240
   resistance, 35
Electroencephalographic monitoring, 121–123
Electroencephalography (EEG), 39–126. *See also specific conditions; specific rhythms*
   abnormal patterns, 43, 44, 85–106
      asymmetry, 113
      background, 113
      definition, 85
      dysmaturation, 112–113
      epileptiform, 45–47, 69
      focal loss, 94
      focal vs. generalized, 90
      frequency composition, 90
      neonatal, 112–115
      periodic patterns, 95, 101–105
      slowing, 43, 44, 65, 85–86, 90–94
      spikes/sharp waves, 43, 45–46, 86–89, 94–105
   activation methods, 58–59, 69, 81–84
   artifacts, 60, 75–81, 84
      60-Hz interference, 78–79
      electrocardiogram artifact, 80, 120, 241
      electrode lead movement, 78–79
      eye movement potentials, 75–76
      eye opening, 76–77
      gel interference, 79
      glossokinetic, 77–78
      machine artifact, 79–80
      muscle artifact, 77
      pulse, 80–81
   basic principles, 41–64
   biological calibration, 57
   brain death, 113, 117–120
   cortical potentials, 41–42
   drowsiness, 66
   duration, 58
   evoked potentials (EPs) and, 197
   face sheets, 55–56
   interpretation, 62
      inpatient EEG, 122
      neonates, 108–109
   isoelectric, 113
   laboratory, 60–61, 121
   monitoring, 121–123
   neonatal. *See* Neonatal EEG
   nocturnal polysomnography, 240, 243

Electroencephalography (EEG)—continued
  normal patterns, 43–45, 65–84
    adults, 65–69
    children, 69–71
    neonates, 109–112
    transients/variants, 71–75, 85
  pen pressure/damping, 48, 57
  physiology, 41–47
  quantitative, 123–125
    brain mapping, 124–125
    method, 123–124
    power spectral analysis, 124
    spike detection, 124
  record keeping, 62–63
  report, 61–62
  rhythm generation, 41–43
  routine, 55–59
  scalp potentials, 11–12, 42–43
  sensitivity, 58
  sleep. See Sleep
  special studies, 117–128
  square-wave calibration, 56–57
  technical aspects, 47–63
    amplifiers. See Amplifiers
    brain death determination, 120
    digital vs. analog, 47–49, 63
    electrodes. See Electrodes
    equipment, 47–50
    filters. See Filters
    monitoring EEG, 121–123
    montages. See Montage
    neonatal EEG, 107–108
    number of channels, 49
    quantitative EEG, 123–124
    reading stations, 50
  telephone transmission, 59–60
  waking, 65–66, 69
Electromotive force (EMF), 17, 19, 32
Electromyography (EMG)
  abnormal, 148–151
    common diagnoses, 130, 149
    complex repetitive discharge (CRD), 149, 150
    differential diagnoses, 150
    fasciculation potentials, 131, 149, 150
    fibrillation potentials, 132, 149
    motor unit potentials (MUPs), 149, 150, 151
    myokymia, 149, 150
    myotonia, 149, 150
    positive sharp waves, 132, 149
    recruitment defects, 149, 151
  basic principles, 144–151
  clinical approaches, 161–166
    classification/identification, 161
    individual muscles, 159
  clinical correlations, 162
  EEG contaminant, 77
  equipment, 134
    electrodes, 144
  giant potentials, 132
  indications, 129–130
  insertional activity, 145, 148, 150–151
  interference patterns, 146
  interpretation, 148–151
  maximal contraction, 145–146, 151
  methodology, 144–148
    electrode positioning, 145, 146
    muscles chosen, 147–148
    single motor unit activation, 145
  myopathies, 133–134, 149, 181–186
  needle, 129, 144
    positioning, 145, 146
  neonatal, 111
  neurodegeneration, 131–132
  neuromuscular junction defects, 133, 152–153, 187–192
  neuropathies, 132–133, 149, 151, 153–159, 167–180
  normal findings, 147, 148
  options, 129
  parameters, 147
  polyphasic potentials, 132
  resting activity, 145, 148–150
  single-fiber recording (SFEMG), 152–153
  sleep studies, 241, 242, 244
    REM sleep, 236
  submental EMG, 241
  temperature effects, 140
Electronic conduction, 6–7
Electronics, 15–38
Electrons, 15, 16. See also Current
Electro-oculogram (EOG), nocturnal polysomnography, 240
Encephalitis, 88, 103
Encephalopathy, EEG, 103
  anoxic, 88, 103
  slow alpha rhythm, 65, 74
  slow delta rhythm (waking), 92
  slow PDR, 85, 91
  slow theta rhythm (waking), 92
  spikes, 88
Endocrine myopathies, 184
Entrapment neuropathies, 153, 154, 155, 156, 173–179

Ephapse, 7
Epilepsy. *See also* Epileptiform activity; Seizures
  absence, 59, 88
  benign focal of childhood, 96–97
    occipital, 88, 95, 97
    rolandic, 88, 95, 96–97
  juvenile myoclonic, 88
  myoclonic, 228
  SEPs, 228
Epileptiform activity, 45–47. *See also* Epilepsy; Seizures
  BETS, 73
  focal, 114
  focal loss, 94
  interpretation, 122
  multifocal, 114–115
  neonatal, 113–115
    alpha rhythmic, 114
  paroxysmal depolarization shift (PDS), 45, 46–47
  periodic, 95, 101–105
    PLEDs, 101–103
  pseudo-beta-alpha-theta-delta discharge, 115
  rhythmicity generation, 46
  rhythmicity termination, 47
  sleep, 69
  slowing, 94
  spikes and sharp waves, 43, 45–46, 86–89, 95–97
    topographic map, 46, 124–125
Equilibrium potential, 4–5
Equipment
  electroencephalography (EEG), 47–50
    cost, 49
    portability, 50
    purchase, 49–50
    service, 50
  electromyography (EMG), 134, 144
  evoked potentials (EPs), 196–197
    minimal requirements, 197
  nerve conduction studies (NCS), 134, 135, 138
    filter frequency responses, 33
    patient interface, 34–36
Erb's palsy, 177
Erb's point, 222
Event-related potentials. *See* Evoked potentials (EPs)
Evoked potentials (EPs), 195–200
  abnormal, 196, 198–199
  analysis time, 198
  artifacts, 197–198
  auditory (brainstem). *See* Brainstem auditory-evoked potentials (BAEPs)
  averaging, 197
  clinical indications, 195–196
  data expression/manipulation, 199
  display, 198
  EEG and, 197
  equipment, 196–197
  general principles, 196–200
  maturation changes, 199
  neural generators, 196
  neurological localization, 195
  normal, 196, 198–199
  physiology, 196
  replications, 198
  reports, 199, 200
  safe vs. unsafe practice, 195
  signal acquisition, 197–198
  somatosensory. *See* Somatosensory-evoked potentials (SEPs)
  stimulus/response, 197, 201–204, 211–213, 222–223
  visual. *See* Visual-evoked potentials (VEPs)
Excitatory postsynaptic potentials (EPSP), 9, 10
Extracellular fluid, 3
Extrafusal muscle fibers, 130
Eye lead placement, 76
Eye movement
  EEG potentials, 75–76, 240
  eye lead placement, 76
  fixation, pattern-reversal VEP, 201, 204
  neonatal EEG, 107
  REM sleep, 236
Eye opening, 76–77
  alpha rhythm suppression, 44

Face sheets, routine EEG, 55–56
Facial nerve palsy (Bell's palsy), 174
Facioscapulohumeral muscular dystrophy (FSH), 183
Farads, 19
Fasciculations, 131, 149, 150
Fatigue, clinical evaluation, 162
Femoral nerve, 158, 174
Fiber density, single-fiber EMG, 152–153
Fibrillation potentials, 132, 149
Field potentials, 12–14
  bipolar, 13, 14
  dipole formation, 42

Field potentials—*continued*
  extracellular, 46
  far-field, 12, 13, 14
  mixed, 12
  near-field, 12–13
  paroxysmal depolarization shift, 46–47
  positive, 41–42
  unipolar, 12
Fields, 17
Filters, 31–32
  60-Hz (notch) filter, 31, 32, 58
  active, 33
  analog, 47
  calibration, 56–57
  default settings, 33
  digital, 33, 47
  frequency response, 33–34
  high-frequency filter (HFF), 31, 32, 56–57, 58
  low-frequency filter (LFF), 31, 32, 56–57, 58
  passive, 33
  standard settings, 58
FIRDA, 45, 85, 93–94
Flash VEP, 201, 202, 209
Foot drop, clinical evaluation, 163
Forward-biased diodes, 23, 24
14-and-6 positive spikes, 71, 73, 89, 94, 101
Frequency response, 33–34
Friedrich's ataxia
  NCS/EMG, 173
  SEPs, 228
Frontal intermittent rhythmic delta activity (FIRDA), 45, 85, 93–94
F-wave study, 129, 141, 171

Gain, digital EEG, 48
Gallium, 22
Gel fixation of electrodes, 35, 50, 51
Generalized seizures, EEG, 88, 96
Generator potential, 5–6, 7
Giant potentials, 132
Glossokinetic artifact, EEG, 77–78
Goldman constant field equation, 5
  generator potentials, 6
Guillain–Barré syndrome, 132–133, 171
  NCS/EMG, 163, 164
  SEPs, 228

Half-field pattern-reversal VEP, 201, 204, 205
Head injury
  EEG, 92
  SEP, 228
Head measurement, 53
Hearing loss, 219–220
Hemispherectomy, SEPs, 228
Hereditary motor sensory neuropathy
  type I (Charcot–Marie–Tooth), 170, 172
  type II, 170, 172
  type III (Dejerine–Sottas), 170, 172
  type IV (Refsum's disease), 170, 172
  type V, 173
Hereditary neuropathies, 170, 172–173
Hereditary pressure-sensitive palsy, SEPs, 228
Herpes simplex encephalitis, EEG, 103
  PLEDs, 103
  spikes, 88
High-frequency filter (HFF), 31, 32
  square-wave calibration, 56–57
  standard settings, 58
High-pass filter. *See* Low-frequency filter (LFF)
H-reflex, 129, 141–142
Hyperkalemic periodic paralysis, 184, 186
Hyperthyroidism
  beta rhythm, 44
  SEPs, 228
Hyperventilation
  contraindications, 59, 84
  EEG activation, 59, 84
  three-per-second spike–wave, 99
Hypokalemic periodic paralysis, 184
Hypothermia, brain death vs., 117, 118
Hysarrhythmia, 101, 102
  interpretation, 98

Impedance, 25, 35, 59
Inclusion body myositis, 183
Inductance, 78
Induction, definition, 20
Inductors, 20–21, 22
Inertial distortion, 48
Inflammatory myopathies, 183
Inhibitory postsynaptic potentials (IPSP), 10
International Federation of Societies for EEG and Clinical Neurophysiology, 52
Interossus syndromes, 175, 177
Intracellular fluid, 3
Intrafusal muscle fibers, 130
Ion channels, 3, 4

Ion pumps, 3, 4
Isoelectric EEG, brain death, 113

Jitter, single-fiber EMG, 152
Juvenile myoclonic epilepsy, 88

K complex, 67
Kirchhoff's current law, 27
Kirchhoff's voltage law, 27–28

Laboratory information, EEG reports, 61
Lambda waves, 71, 72
Lambert–Eaton (myasthenic) syndrome, 133, 159, 188–189
Large-fiber neuropathy, 170
Lennox–Gastaut syndrome, EEG, 88, 100
Leprosy, neuropathy, 180
Leukodystrophies, SEPs, 228
Ligament of Struthers, 175
Limb-girdle dystrophy, 182
Lipid bilayer, insulation, 3
Longitudinal bipolar (LB) montage, 53, 54
 neonates, 108
Low-frequency filter (LFF), 31, 32
 square-wave calibration, 56–57
 standard settings, 58
Low-pass filter. See High-frequency filter (HFF)
Low-voltage background, neonatal EEG, 113
Lumbar plexopathy, 163
Lumbar radiculopathy, 163

Machine artifact, 79–80
Magnetic fields, 17
 electrode leads, 78
 inductors, 20–21
 right-hand rule, 20, 21
Mechanical distortion, digital vs. analog EEG, 48–49
Median nerve
 anatomy, 153
 compression, 175
 Erb's point, 222
 evaluation, 153–154
 nerve conduction studies (NCS), 135, 139, 153
 neuropathy, 153–154, 173–175
 SEP analysis, 222–225
  abnormalities, 225
  indications, 224–225
  method, 222–223
  N9 wave, 222, 223
  N20 wave, 223

P14 wave, 223
Medical Consultants on the Diagnosis of Death (1981), 117, 119, 219
Membrane potential, 3–5
 action potentials, 7
 components, 3
 equilibrium potential, 4–5
 resting, 5
Membrane proteins, 3
Membranes, 3, 4
Meralgia paresthetica, 163
Metabolic disorders
 brain death vs., 117
 myopathies, 184–185
 neuropathies, 171–172
Miniature end-plate potentials (MEPPs), 152
Minimata disease, SEPs, 228
Mitochondrial myopathies, 184
Mittens, 71, 75
Mixed sleep apnea, 246–247
Mononeuropathies, 170, 173–179
 diabetic, 170–171
Mononeuropathy multiplex, 170, 179–180
Montage, 53–55
 bipolar, 53, 54, 55
 brain death confirmation, 119
 digital vs. analog EEG, 48
 Guidelines, 54–55
 inpatient monitoring, 121–122
 neonatal EEG, 107–108
 nocturnal polysomnography, 240
 quantitative EEG, 123–124
 referential, 53–54, 55, 80
 SEPs, 222–223, 226
 VEPs, 204–205
Motoneuron disease (MND), 163
Motoneurons
 axon diameter, 130–131
 continuous activation, 185
 degeneration, 131–132
  NCS/EMG, 162
 dysfunction, 132
 F-waves, 141
 neuromuscular transmission, 10, 130
Motor conduction studies, 129, 135–138, 136
 blink reflex studies, 142–144
 CMAP, 132, 136, 137, 142, 152
 electrodes, 135
 F-wave study, 141
 H-reflex, 141–142
 interpretation, 137–138

Motor conduction studies—*continued*
  measurements, 137
  median nerve, 135
  methods, 135–136
  NCV abnormalities, 137–138
    conduction block, 138
    increased distal latency, 137
    relative slowing, 137–138
    slow motor NCV, 137
  normal findings, 138
  parameters, 136
  stimulus characteristics, 135–136
Motor function, 130–131, 221
Motor units, 130
  EMG potentials, 145, 149, 150, 151
  single motor unit activation, 145
Movement
  EEG artifact, 78–79, 84
  nocturnal polysomnography, 241–242
Multifocal motor neuropathy (MMN), 171
Multiple sclerosis (MS), evoked
      potentials, 195
  BAEPs, 219
  SEPs, 228, 229
  VEPs, 207–208
Multiple sleep latency test (MSLT), 237, 244–245
Muscle
  biopsy, 184
  contraction, 10, 11, 130–131
    maximal, 145, 151
  denervation, 132
  disorders, 186
    myopathy. *See* Myopathies
  electrical properties, 10
  EMG study. *See* Electromyography (EMG)
  evaluation, 159
  fibers, 11, 130, 131
    continuous activity, 185–186
  innervation, 153–159
  neuromuscular transmission. *See* Neuromuscular junction (NMJ)
  physiology, 10
Muscle artifact, EEG, 77
Muscular dystrophies, 181–183
  Becker's, 181–182
  clinical findings, 164, 182
  Duchenne, 181–182
  facioscapulohumeral, 183
  limb-girdle, 182
  myotonic, 182–183
  scapuloperoneal, 183
Mu waves, 71, 72

Myasthenia gravis, 159, 187–188
  clinical findings, 164, 187
  neonatal, 190–191
  ocular, 188
  repetitive stimulation, 188, 189
Myelinated nerves
  action potential propagation, 8
  myelin dysfunction, 132–133
Myelin sheath, 8
Myelitis, transverse, 228
Myoclonic epilepsy
  juvenile, spikes, 88
  SEPs, 228
Myoclonic jerks, PLEDs, 103
Myokymia, 149, 150
Myopathies, 133–134. *See also specific disorders*
  clinical evaluation, 162
  clinical findings, 164, 182, 183, 184
  continuous fiber activity syndromes, 185–186
  inflammatory, 183
  metabolic, 184–185
  muscular dystrophies, 181–183
  NCS/EMG, 133–134, 149, 151, 162, 181–186
Myotonia, 149, 150, 182–183, 185–186
Myotonia congenita (Thomsen disease), 184, 186
Myotonic dystrophy, 182–183
  NCS/EMG, 182–183
  SEP, 228

Narcolepsy, 246
  REM-onset sleep, 68, 245
Needle electrodes, 35, 51, 144
Negative (N-) doping, 22
Neonatal EEG, 107–116
  abnormal, 112–115
    background, 113
    epileptiform, 113–115
  interpretation guidelines, 108–109
  maturation, 110–112
    abnormal (dysmature), 112–113
    bursts, 110
    conceptual age 22–29 weeks, 110
    conceptual age 29–31 weeks, 110–111
    conceptual age 32–34 weeks, 111
    conceptual age 34–37 weeks, 111–112
    conceptual age 38–40 weeks, 112
    delta brushes, 111

normal, 109–112
premature infants, 111
sleep, 109
technical aspects, 107–108
tracé alternant (TA), 109, 112
tracé discontinu (TD), 110
Neonatal EPs, 199
  auditory, 216, 218
Neonatal myasthenia, 190–191
Neonates
  conceptual age, 109
  EEG. *See* Neonatal EEG
  EPs, 199, 216, 218
  myasthenic syndromes, 190–191
  physiologic state, 109
  sleep–wake cycle, 109
Nernst equation, 4
Nerve(s)
  conduction
    action potential. *See* Action potentials
    afferent volley, 221
    block, 133, 138, 167
    electrical activity, 10–12
    electronic, 6–7
    motor, 130–131
    sensory, 131
    studies. *See* Nerve conduction studies (NCS)
    velocity (NCV), 134, 137–138, 140–141, 168–180
  connections, 11–12
  synapses, 11
  disorders, 131–134
  evaluation, 153–159
  physiology, 3–14
Nerve bundles, 11, 196
Nerve conduction studies (NCS)
  amplitude, 134
  artifacts, 136
  basic principles, 134–144
  blink reflex, 142–144
  clinical approaches, 161–166
    classification/identification, 161
    individual nerves, 153–159
  clinical correlations, 162
  common diagnoses, 130
  conduction velocity (NCV), 134
    abnormalities, 137–138, 140, 168–180
    age effects, 140
    calculation, 137, 140
    temperature effects, 140–141
  dispersion, 133

  equipment, 134
    electrodes, 135, 138
    filter frequency responses, 33
  F-wave, 141, 171
  H-reflex, 141–142
  indications, 129–130
  interpretation, 137–140, 143
  methods, 135–136, 138–139, 142–144
  parameters, 136
  positioning electrodes, 135, 138–139
  stimulus characteristics, 135–136
  motor. *See* Motor conduction studies
  muscle disorders, 186
  myopathies, 133–134, 181–186
  neuron degeneration, 132
  neuropathies, 132–133, 153–159, 167–180
  NMJ defects, 132, 133, 152, 187–192
  normal findings, 138, 144
  routine vs. nonroutine, 129
  sensory. *See* Sensory conduction studies
Neurodegenerative disorders, 131–132
Neuromuscular disorders, 129–192. *See also specific disorders*
  classification/identification, 161
  clinical approaches, 153–159, 161–166
    individual muscles, 159
    individual nerves, 153–159
    neuromuscular transmission, 159
  evaluation, 161–166
    basic principles, 129–159
    clinical, 162–164
    common findings, 164–165
    conduction studies. *See* Nerve conduction studies (NCS)
    electromyography. *See* Electromyography (EMG)
  myopathy. *See* Myopathies
  neuropathy. *See* Neuropathies
  physiology, 131–134
    neuron cell body dysfunction, 131–132
    peripheral axon dysfunction, 132
    peripheral myelin dysfunction, 132–133
  transmission defects, 133, 152, 159, 162, 187–192

Neuromuscular junction (NMJ), 10, 151–153
  acetylcholine. *See* Acetylcholine
  brain death vs. blockade, 117, 118
  clinical approaches, 159, 161–166
  defects, 133, 152, 159, 162, 187–192
    autoantibodies, 187
    botulism, 159, 189–190
    clinical features, 188
    Lambert–Eaton (myasthenic) syndrome, 133, 159, 188–189
    myasthenia gravis, 159, 164, 187–188
    neonatal myasthenia, 190–191
  evaluation, 159
  miniature end-plate potentials (MEPPs), 152
  motor function, 130–131
    motoneurons, 10, 130
  physiology, 130–134
    abnormal, 131–134
    normal, 130–131
  repetitive stimulation, 151–152, 188, 189
    CMAP amplitude, 152
  sensory function, 131
Neuromuscular transmission. *See* Neuromuscular junction (NMJ)
Neuromyotonia, 185–186
Neuronal degeneration, 131–132, 162, 168–169
Neuropathies, 167–180. *See also specific disorders*
  common, 167–168
  demyelinating, 132–133, 162, 171
  entrapment, 153, 154, 155, 156, 173–179
  evaluation, 153–159
    clinical, 162, 163, 164
    common findings, 164–165
    nerve conduction/EMG, 149, 151, 162, 167–180
  metabolic/toxic, 171–172
  mononeuropathies, 170, 173–179
  mononeuropathy multiplex, 170, 179–180
  multifocal motor neuropathy (MMN), 171
  neuronal degeneration, 131–132, 162, 168–169
  peripheral, 132, 153, 163, 167, 229
  polyneuropathies, 133, 162, 169–173
  radiculopathy, 162, 163, 164, 179, 180
Neurotoxins, 189–190
Neurotransmitters, 9–10. *See also* Action potentials; Synapses
  generator potentials, 6
  release, 9
Neutrons, 15
Nocturnal polysomnography, 237, 239–244
Node, definition, 27
Nonconductors, 15, 22
Notch (60-Hz) filter, 31, 32, 58
Nuclei, evoked potential generation, 196

Obstructive sleep apnea, 246–247
Occipital epilepsy, EEG spikes, 88, 95, 97
Occipital intermittent rhythmic delta activity, 45, 93–94
Ocular disorders, VEPs, 209
Ocular myasthenia, 188
Ohms, 18
Ohm's law, 25–26
Ophthalmological evaluation, 195
Optic neuritis, evoked potentials, 207, 208
Orbital shells, 15, 16
Overshoot, 48

Panencephalitis, subacute sclerosing (SSPE), 88, 104
Paramyotonia congenita (of Eulenberg), 186
Parapareisis, 195
Parenchymal destruction, PLEDs, 102–103
Parietal lesion, 228
Parietal sharp waves, 95
Paroxysmal depolarization shift (PDS), 45, 46–47
Passive filters, 33
Patient–equipment interface, 34–36
Patient information, EEG reports, 61
Pattern-reversal VEP, 201, 202–204, 209
Peak potential, 8
Pediatric BAEPs, 216–218
Pen pressure, EEG, 57
Perinatal asphyxia
  EEG, 115
  SEP, 228
Periodic lateralized epileptiform discharges (PLEDs), 101–103
Periodic paralysis, 184, 186

Peripheral nerve(s). *See also specific nerves*
  cortical projections, 222
  demyelination, 132–133
  neuropathies, 132, 153, 163, 167, 229
Permeability, definition, 5
Peroneal nerve, 155–157
  anatomy, 157
  conduction studies, 155–156
  entrapment neuropathy, 156, 174, 177–178
  palsy, 163, 164
Persistent vegetative state, SEPs, 228
Photic stimulation
  EEG activation, 59, 81–83
  method, 81
  newborns, 107
  normal responses, 81–82
    driving response, 82
    visual-evoked, 81, 82
  photoconvulsive response, 83
  photomyoclonic response, 83
Photoconvulsive response, 83
Photomyoclonic response, 83
Physics, 15–38
Physiology, 3–14
  EEG, 41–47
  evoked potentials, 196
  neuromuscular transmission, 130–134
  sleep, 233–238
PIRDA, 45, 93–94
Piriformis syndrome, 179
Plexopathy, 162, 163
Poliomyelitis, 168, 169
Polyarteritis nodosa, neuropathy, 180
Polymorphic delta activity (PDA), 92–93
  FIRDA vs., 94
Polymyositis, 164, 183
Polyneuropathies, 133, 162, 169–173
  clinical features, 170
  diabetic, 170–171
  NCS/EMG, 169–173
  SEPs, 228
Polyphasic potentials, 132
Polysomnography (PSG), 231–248. *See also* Sleep; Sleep disorders
  basic principles, 233
  indications, 233, 237
  multiple sleep latency test (MSLT), 237, 244–245

  interpretation, 245
  methods, 244–245
  nocturnal, 237, 239–244
    grading, 243
    interpretation, 243–244
    physiologic measurements, 239–242
    recording protocol, 242
  sleep disorders, 239–248
Positive (P-) doping, 22
Positive occipital sharp transients of sleep (POSTS), 71, 72–73, 89
Positive sharp waves, 132, 149
Posterior dominant rhythm (PDR), 43–44, 235. *See also* Alpha rhythm; Theta rhythm
  interpretation, 91
  maturation, 69
  slowing, 65, 85, 90–91
  subharmonic (normal variant), 91
Posterior intermittent rhythmic delta activity (PIRDA), 45, 93–94
Posterior slow waves of youth, 69
Postpolio syndrome, 169
POSTS, 71, 72–73, 89
Potassium permeability, 4
Potential field, spikes/sharp waves, 87
Power spectrum, 31
  quantitative EEG, 124
Premature infants
  audiometry, 218
  BAEPs, 216
  neonatal EEG, 111
President's Commission for the Study of Ethical Problems in Medicine (brain death), 117, 119, 219
Primary lateral sclerosis, 168
Pronator teres syndrome, 164, 175
Protons, 15
Pseudo-beta-alpha-theta-delta discharge, 115
Pseudomotor cerebri, VEPs, 208
Pseudoseizures, EEG monitoring, 122, 123
Psychomotor theta variant. *See* Rhythmic mid-temporal theta of drowsiness
Pulse artifacts, EEG, 80–81

Quantitative EEG, 123–125
Quiet sleep, neonatal EEG, 109, 110, 111

Radial nerve, 155–156
  anatomy, 156
  conduction studies, 155
  neuropathy, 155, 174, 176–177
    clinical evaluation, 163
    posterior interossus
      syndrome, 177
    spiral groove, 176
Radiculopathy
  NCS/EMG, 162, 163, 164, 179, 180
  SEP, 228
Raphe nucleus, sleep role, 234
Rapid eye movement sleep.
  See REM sleep
Receptor binding, neurotransmitters,
  9–10
Record keeping, EEG reports, 61–63
Rectifiers, 24
Referential montage, 53–54, 55
Refractory periods, 8
Refsum's disease, 170, 172
Regenerative depolarization, 7
REM-onset sleep, 68, 245
REM sleep, 68, 235–236
  progression, 69
Renal failure, SEPs, 228
Repolarization, 8
Reports
  EEG, 61–62
  EP, 199, 200
Resistance, 18–19
  additive, 29
  electrode, 35
  measurement, 18
  Ohm's law, 25–26
Resistor–capacitor (RC)
    circuits, 32–33
Resistors, 18–19
  circuit properties, 28–30
  resistor–capacitor circuits, 32–33
  series/parallel, 28–30
Respiration
  neonatal EEG, 107
  nocturnal polysomnography, 241
Reticular formation, sleep role, 234
Reverse-biased diodes, 23, 24
Reye's syndrome, SEPs, 228
Rhythmicity, 42
Rhythmic mid-temporal theta of
    drowsiness, 71, 74–75
  seizures vs., 74
Right-hand rule, 20
Rolandic epilepsy, EEG spikes, 88, 95,
    96–97

Saturday night palsy, 174
Scalp potentials, 42–43
Scapuloperoneal muscular dystrophy, 183
Schwartz–Jampel syndrome, 185
Sciatic nerve, 158
  conduction studies, 158
  EMG, 158
  neuropathy, 158, 179–180
    clinical evaluation, 163
    piriformis syndrome, 179
    sciatic stretch, 180
Sciatic stretch, 180
Sedatives, beta rhythms, 44, 65
Seizures. See also Epileptiform activity
  absence, 59, 88, 99
  akinetic, 100
  complex partial, 88, 96
  generalized, 88, 96
  interpretation, 122
  neonatal EEG, 113–115
  pseudoseizures, 122, 123
  simple partial, 88, 95–96
  slow waves, 94
  spikes, 88, 95–97, 99, 100
  tonic–clonic, 88, 100
Semiconductors, 21–25
  definition, 15
  diodes, 22–24
  doping, 21, 22
  theory, 23
  transistors, 24–25
Sensorimotor neuropathy, diabetic, 170
Sensorineural hearing loss, 219–220
Sensory conduction studies, 129, 136,
    138–141
  averaging, 139
  electrodes, 138–139
  interpretation, 139–140
  measurements, 137–138
  median nerve, 139
  methods, 138–139
  NCV abnormalities, 138
  nonpathological factors, 140–141
  normal findings, 138
  SNAP, 131, 139, 169
Sensory nerve action potential (SNAP),
    131, 139, 169
Sensory nerves, 131, 162, 221
Silicon, 22
Simple partial seizures, EEG spikes, 88,
    95–96
Single-fiber recording (SFEMG), 152–153
Six-per-second (phantom) spike and
    wave, 89, 100–101

interpretation, 98, 101
Sleep
  anatomy, 234
  blood oxygenation, 241
  CO_2 expiration, 241–242
  disorders. *See* Sleep disorders
  EEG activation, 59, 69
  EEG recording, 44, 45, 66–69, 234.
    See also Polysomnography (PSG)
    children, 69–71
    epileptiform activity, 69, 73
    generalized slowing, 44, 92
    newborns, 109
    nocturnal polysomnography, 240, 244
  efficiency, 243
  electrocardiography (ECG), 241
  electro-oculogram (EOG), 240
  indications for study, 233, 237
  latency test, 237, 244–245
  movement during, 242
  onset definition, 243
  physiology, 233–238
  respiration, 241
  rhythms, 235, 236
    background rhythm, 66
    beta rhythm, 44
    delta rhythm, 45
    K complex, 67
    positive sharp transients, 67
    slowing, 92
    spindles, 67, 70, 71, 75, 236
    theta rhythm, 44
    tracé alternant (TA), 109
    vertex waves, 66–67, 70–71, 236
  stages, 67–68, 234–237. *See also* Waking state
    drowsiness (stages 1A/B), 66, 67, 68, 70, 71, 74–75, 235
    neonatal, 109
    REM sleep, 68, 69, 236–237
    sequence, 68–69, 243
    stage 2 (light), 67, 70, 235, 236
    stage 3 (slow-wave), 67, 236
    stage 4 (slow-wave), 68, 236
  submental EMG, 241
  visual information, 72
Sleep apnea, 246–247
  nocturnal polysomnography, 241
Sleep deprivation, 234
  MSLT, 245
  REM-onset sleep, 68
Sleep disorders, 246–247

classification, 246
narcolepsy, 68, 245, 246
physiology, 233–238
polysomnography, 233, 239–248
  MSLT, 244–245
  nocturnal, 239–244
  REM-onset sleep, 68, 245
  sleep apnea, 241
  sleep latency alteration, 245
Sleep–wake cycle, 234
  neonates, 109
Slow alpha variant rhythm, 71, 73–74
Slow-channel syndrome, 190
Slow motor NCV, 137
Slow rhythms
  abnormal, 43, 44, 45, 85–86, 90–94
    CJD, 105
    dementia, 65
    encephalopathy, 65, 85
    FIRDA, 45, 85
    focal slowing, 86, 90, 92–93
    generalized slowing, 85, 86, 90–92
    neonates, 113
    posterior dominant, 65, 85, 90–91
    regional slowing, 85
    seizures, 94
    sharp wave association, 98
    waking state, 92
  normal
    posterior slow waves of youth, 69
    sleep, 67, 68
    slow alpha variant rhythm, 71, 73–74
Slow-wave sleep, 67, 68, 236
Small-fiber neuropathy, 170
SNAP, 131, 139, 169
Sodium channels
  electronic conduction, 6
  voltage-gated, 7, 130, 132
Sodium–potassium pump, 3, 4
Somatosensory-evoked potentials (SEPs), 201, 221–230
  abnormal, 225, 227, 229
  afferent nerve volley, 221
  analysis time, 198
  basic principles, 221–222
  brain, 221
  brain conduction time (BCT), 224
  clinical correlations, 207, 228–230
  clinical indications, 195
  commonly used nerves, 221
  data, 224–225, 227
  interpretation, 227, 229

Somatosensory-evoked potentials
(SEPs)—*continued*
  median SEP, 222–225
  methods, 222–223, 225–226
  montage, 222–223, 226
  neural generators, 195
  normal, 201
  recording parameters, 222, 226
  spinal cord, 221
  stimulus parameters, 222, 225, 226
  tibial SEP, 225–229
  waveforms
    identification/interpretation,
      223–225, 226–227, 229
    median nerve, 223–224
    origins, 225
Sphenoidal electrodes, 51
Spikes/sharp waves, 43, 45–46, 86–89,
  94–105
  clinical correlations, 87–88, 99–100
  clinical interpretation, 89, 95, 98,
    100–101
  detection, 34
    quantitative EEG, 124
  diffuse (generalized), 90, 98–101
    fast spike–wave complex, 100, 101
    hypsarrhythmia, 98, 101, 102
    phantom, 89, 98, 100–101
    slow spike–wave complex, 100
    slow wave association, 98
    three-per-second spike–wave,
      98–100
  focal, 90, 94–98
    neonates, 114
  misinterpretation, 87
  multifocal, 114–115
  neonates, 113–115
  nonseizure-associated, 89, 97–98
  nonspike vs., 87
  normal variants, 71–75, 89, 94
  potential field, 87
  seizure-associated, 88, 95–97. See
    also Seizures
Spinal artery syndrome, posterior, 228
Spinal cord disorders, SEPs, 221, 222,
  228, 229–230
  cervical data, 227
Spinal muscular atrophy (SMA), 131,
  168–169
Spinal stenosis, clinical
  evaluation, 163
Spindles, 67, 70, 71, 236
Square-wave calibration, 56–57
SREDA, 71, 75, 94, 97

Steroid myopathy (Cushing's
  syndrome), 184
Stiff-man syndrome, 185
Stimulators, NCS, 135–136
Storage, digital EEG, 48
Stray capacitance, 78
Stray inductance, 78
Stroke
  BAEPs, 218–219
  PLED, 102
  SEPs, 218–219
Structural lesions
  EEG, 92, 94
  VEP, 208
Subacute sclerosing panencephalitis
  (SSPE), EEG, 88, 104
Subarachnoid hemorrhage, SEPs, 228
Subclinical rhythmic electrographic
  discharge of adults (SREDA), 71,
  75, 94, 97
Subdural strip electrodes, 51
Subharmonic posterior dominant
  rhythm, 91
Submental EMG, 241
Sural nerve, 157–158
Surface electrodes, 50–51
Synapses, 9, 11
  orientation, 196
Synaptic transmission, 9
  chemical, 9–10
  electrical, 6–7

Talking, glossokinetic artifact, 77–78
Tarsal tunnel syndrome, 163, 164
Task Force for Brain Death in
  Children, 119
Telephone transmission EEG, 59–60
Temporal sharp waves, 95
10–20 Electrode Placement System, 52
Tetanus, 185
Thalamocortical efferents, 41
Theta rhythm, 44–45. See also Posterior
  dominant rhythm (PDR)
  rhythmic mid-temporal theta of
    drowsiness, 71, 74–75
  sleep, 44
  slowing, 92, 105
  vascular disease, 45
  waking state, 92
Thomsen disease (myotonia
  congenita), 184, 186
Thoracic outlet syndrome
  NCS/EMG, 164
  SEPs, 228

Three-per-second spike–wave, 98–100
  appearance, 98–99
  clinical correlations, 99–100
  hyperventilation, 99
  interpretation, 98
Tibial nerve, 156–158
  anatomy, 158
  entrapment neuropathy, 178
  SEP analysis, 225–229
    abnormalities, 227, 229
    data, 227
    LP (lumbar potential), 226, 227, 229
    method, 225–226
    N34 wave, 227
    P37 wave, 227
Tight junctions, electronic conduction, 7
Time constant (TC), 32, 58
Time measurement, nocturnal polysomnography, 242
Tones, auditory-evoked potentials, 212
Tonic–clonic seizures, EEG, 88, 100
Topographic maps. *See* Brain mapping
Tourette's syndrome, SEPs, 228
Toxic neuropathies, 171–172
Tracé alternant (TA), 109, 112
Tracé discontinu (TD), 110
Transcranial doppler (TCD), brain death, 118
Transients (EEG). *See also specific patterns*
  abnormal variants, 43, 45–46, 86–89
  definition, 86
  normal variants, 71–75, 89, 94
Transistors, 24–25
Transverse myelitis, SEPs, 228
Tumors
  BAEPs, 218
  EEG, 92, 94
  VEPs, 208

Ulnar nerve, 154–155
  anatomy, 155
  conduction studies, 154
  EMG, 154
  neuropathy, 154, 174, 175–176
    clinical evaluation, 164
    common findings, 164–165
    palmar lesion, 175
Unmyelinated nerves, action potential propagation, 8

Vascular disease
  PDR, 91
  theta rhythm, 45, 92

Vertex waves (sleep), 89, 236
  adult, 66–67
  blunted morphology, 70
  children, 70–71
  clustering, 70
  K complex, 67
  mittens, 75
Viral myositis, 183
Visual-evoked potentials (VEPs), 201–210
  abnormal, 207
  clinical correlations, 195, 207–209
  clinical indications, 195
  flash VEP, 201, 202, 209
  interpretation, 205–207
  methods, 201–205
    analysis time, 198
    electrode placement, 204–205
    montage, 204–205
    recording parameters, 203, 205
    stimulus, 201–204
  N75, 205–206
    inversion, 207
  N145, 205–206
    inversion, 207
  neural generators, 196
  normal, 207
  normal photic response, 81, 82
  P100, 205–206
    abnormal, 208
    bifid pattern, 206
  pattern-reversal VEP, 201, 202–204
    checkerboards, 202–203
    clinical use, 209
    contrast, 204
    fixation, 201, 204
    half-field, 201, 204, 205
    luminance, 203–204
    reversal rate, 203
    stimulus field, 203
  waveforms
    identification, 205–206
    variant, 206–207
Visual loss, VEPs, 195, 209
Vitamin B_{12} deficiency, SEPs, 229
Voltage drop, 18, 28, 35
Voltage-gated sodium channels, 7, 130
  neurodegeneration, 132
Voltage source, 18, 28

Waking state, EEG, 234–235
  adults, 65–66
  anterior cerebral activity, 65

Waking state—*continued*
  children, 69
  neonates, 109
  posterior cerebral activity, 65–66
  slowing, 92
  three-per-second spikes, 99

Weakness
  clinical evaluation, 162
  polyneuropathy, 169
White noise, auditory-evoked potentials, 212
Wicket spikes, 71, 73, 89, 94